Future Communication Systems Using Artificial Intelligence, Internet of Things and Data Science

Future Communication Systems Using Artificial Intelligence, Internet of Things and Data Science mainly focuses on the techniques of artificial intelligence (AI), Internet of Things (IoT) and data science for future communications systems.

The goal of AI, IoT and data science for future communications systems is to create a venue for industry and academics to collaborate on the development of network and system solutions based on data science, AI and IoT. Recent breakthroughs in IoT, mobile and fixed communications and computation have paved the way for a data-centric society of the future. New applications are increasingly reliant on machine-to-machine connections, resulting in unusual workloads and the need for more efficient and dependable infrastructures. Such a wide range of traffic workloads and applications will necessitate dynamic and highly adaptive network environments capable of self-optimization for the task at hand while ensuring high dependability and ultra-low latency.

Networking devices, sensors, agents, meters and smart vehicles/systems generate massive amounts of data, necessitating new levels of security, performance and dependability. Such complications necessitate the development of new tools and approaches for providing successful services, management and operation. Predictive network analytics will play a critical role in insight generation, process automation required for adapting and scaling to new demands, resolving issues before they impact operational performance (e.g., preventing network failures and anticipating capacity requirements) and overall network decision-making. To increase user experience and service quality, data mining and analytic techniques for inferring quality of experience (QoE) signals are required.

AI, IoT, machine learning, reinforcement learning and network data analytics innovations open new possibilities in areas such as channel modeling and estimation, cognitive communications, interference alignment, mobility management, resource allocation, network control and management, network tomography, multi-agent systems and network ultra-broadband deployment prioritization. These new analytic platforms will aid in the transformation of our networks and user experience. Future networks will enable unparalleled automation and optimization by intelligently gathering, analyzing, learning and controlling huge volumes of information.

Future Communication Systems Using Artificial Intelligence, Internet of Things and Data Science

Edited by
Dr Inam Ullah, Dr Inam Ullah Khan, Dr Mariya Ouaissa,
Dr Mariyam Ouaissa and Dr Salma El Hajjami

CRC Press
Taylor & Francis Group
Boca Raton London New York

CRC Press is an imprint of the
Taylor & Francis Group, an **informa** business

Designed cover image: Shutterstock Images

First edition published 2024
by CRC Press
2385 NW Executive Center Drive, Suite 320, Boca Raton FL 33431

and by CRC Press
4 Park Square, Milton Park, Abingdon, Oxon, OX14 4RN

CRC Press is an imprint of Taylor & Francis Group, LLC

ISBN: 9781032632032 (hbk)
ISBN: 9781032648279 (pbk)
ISBN: 9781032648309 (ebk)

DOI: 10.1201/9781032648309

Typeset in Minion
by codeMantra

Contents

Preface, vii

Editors, ix

Contributors, xiii

SECTION I **AIoT and Challenges for Future Communication Systems**

CHAPTER 1 ▪ Artificial Intelligence in the Internet of Things, Recent
Challenges and Future Prospects   3

FAZAL WAHAB, MOSA KHAN, INAM ULLAH AND YUNING TAO

CHAPTER 2 ▪ Artificial Intelligence (AI)-Powered Internet of Things (IoT):
Smartening Up IoT   18

AMAL SAIF AND QASEM ABU AL-HAIJA

CHAPTER 3 ▪ Bridging the Future: The Confluence of Internet of Things
and Artificial Intelligence in Communication System   30

RAHIM KHAN, XUEFEI MA, SHER TAJ, HINA HASSAN, INAM ULLAH,
ABDULLAH ALWABLI, YUNING TAO AND HABIB ULLAH

SECTION II **The Future of Data Analytics in Communication**

CHAPTER 4 ▪ The Future of Artificial Intelligence in Communication   57

SADAF HUSSAIN, TANWEER SOHAIL, RABIA AFZAAL AND MUHAMMAD ADNAN KHAN

CHAPTER 5 ▪ Data Science Meets Intelligent Internet of Things   73

INAM ULLAH, IJAZ AHAMD, MUHAMMAD SHAHID ANWAR, YUNING TAO
AND MUHAMMAD SHAFIQ

CHAPTER 6 ▪ Data Science and Big Data Analytics   92

FAISAL REHMAN, MUHAMMAD MUNEER, MUHAMMAD HAMZA SAJJAD AND NAVEED RIAZ

CHAPTER 7 ■ Artificial Intelligence and Machine Learning with Cyber Ethics for the Future World 110

WASSWA SHAFIK

SECTION III **IoT-Based Techniques for Smart Future Architectures**

CHAPTER 8 ■ Internet of Metaverse Things (IoMT): Applications, Technology Challenges and Security Consideration 133

MUHAMMAD SHAHID ANWAR, WADEE ALHALABI, AHYOUNG CHOI, INAM ULLAH AND AHAD ALHUDALI

CHAPTER 9 ■ Social Internet of Things (SIoT): Recent Trends and Its Applications 159

IRSHAD KHALIL, ADNAN KHALIL, INAM ULLAH, YUNING TAO, IJAZ KHAN, SHAHZAD ASHRAF AND WALEED M. ISMAEL

CHAPTER 10 ■ The Role of Software Defined Internet of Things (SDIoT) in Cloud Computing 193

HANIA BATOOL, ADILA MEHDI AND AHTHASHAM SAJID

CHAPTER 11 ■ Internet of Vehicles (IoV): Challenges, Threats and Routing Protocols 205

MARIYA OUAISSA, MARIYAM OUAISSA, SOUKAYNA RIFFI BOUALAM, ZAKARIA BOULOUARD, INAM ULLAH KHAN AND SARAH EL HIMER

CHAPTER 12 ■ Edge Computing in the Digital Era: The Nexus of 5G, IoT and a Seamless Digital Future 213

ZAHID RASHEED, YONG-KUI MA, INAM ULLAH, YUNING TAO, IJAZ KHAN, HABIB KHAN AND MUHAMMAD SHAFIQ

INDEX, 235

Preface

TECHNOLOGICAL ADVANCEMENT HAS CHANGED the overall world dynamics. Integration of artificial intelligence (AI), Internet of Things (IoT) and data science has transformed the landscape of communication systems. Future communication systems will be mainly based on data. New technologies will easily connect and exchange information from one place to another.

AI and machine learning are used to improve the efficiency of intelligence communication systems. With the help of machine learning techniques, predictive analysis can be made possible, while AI-based techniques will easily redefine communication systems. Also, cyber ethics will play a pivotal role during human-machine interaction.

Interconnectivity can be made possible through the IoT. Intelligent communication systems will facilitate real-time applications within smart homes and cities, which will lead to industrial automation.

Data science is basically utilized to extract meaningful information generated by Artificial Intelligence of Things (AIoT) devices. This will open new opportunities for researchers, engineers and scientists. The merger of AI, IoT and data science will enhance communication systems.

This book provides detailed information about future communication systems. Also, it explains the relationship between AI, IoT and data science. Each chapter presents technological advancements, solutions, ethical considerations and opportunities. This book presents a detailed guide for researchers and practitioners about future communication systems.

This edited book proposes comprehensive knowledge about AI, IoT, data science, social IoT, internet of metaverse things, software defined IoT, cloud computing, routing protocols, 5G, digital computing, threats, internet of vehicles, machine learning and future of communication systems. We invite readers all around the world to explore the limitless possibilities and impact of advanced technologies. The technological revolution is reshaping the entire world through intelligent and connected communication systems.

This book will be divided into three sections. The first section will introduce AI and IoT as actors that can help address the challenges of future communication systems. The second section will go in depth and shed light on the future of data analytics in communication, and the last section will explore IoT techniques applied in smart future architectures.

SECTION I: AIoT AND CHALLENGES FOR
FUTURE COMMUNICATION SYSTEMS

In the opening section, we set the stage. Chapter 1 provides a comprehensive study that explores the integration of AI and IoT. AI plays an important role in the advancement of societies. Data-centric decision-making will improve our daily lives. Chapter 2 focuses on AI and IoT applications that are reshaping industries. This integration of technologies will boost the healthcare and transportation industry. This transformation will lead to smart cities. With the help of AI-based tools, precision agriculture, networked environment, resource optimization and ethical considerations need to be regulated. Chapter 3 presents the merger of IoT with AI. Basic concepts of IoT, challenges and real-time applications are discussed. This chapter provides an understanding of the synergistic potential of IoT and AI in shaping the future of communication systems.

SECTION II: THE FUTURE OF DATA ANALYTICS IN COMMUNICATION

Our journey then takes a deep dive into the future of data analytics in communication. Chapter 4 illustrates the future of AI in communication. Privacy, security, transparency and accountability are discussed in detail. AI's impact on human-computer interaction is briefly explained. AI-enabled chatbots, virtual assistants, and language models are discussed. Chapter 5 aims to investigate the potential of convergence to disrupt industries, build smart cities and nurture a more linked and efficient future through real-world examples. Basically, this chapter explains well the concept of data science with IoT. Chapter 6 covers previous research in big data analytics aimed at exploring new patterns in this field to solve academic and business problems, as well as addressing current obstacles. Chapter 7 demonstrates the significance of integrating cyber ethics into the fabric of MAI development, enabling a future world where these technologies serve as tools for positive societal transformation.

SECTION III: IoT-BASED TECHNIQUES FOR
SMART FUTURE ARCHITECTURES

IoT techniques applied in smart architectures emerge as focal points. Chapter 8 provides an in-depth overview of common Internet of Metaverse Things (IoMT) applications and applications of metaverse in different sectors such as healthcare and therapy, education and training, gaming and entertainment, smart city, estate, retail, e-commerce, socialization, work and collaboration, advertising and marketing. In Chapter 9, a novel idea is introduced which is named as social IoT. Related trends and applications of SIoT are discussed. Chapter 10 is basically a survey paper, where the history of SDN and its integration into cloud computing systems are discussed. Also, this chapter focuses on the architecture of SDN-enabled IoT cloud networks, and its related applications are discussed. Chapter 11 explores the challenges and opportunities of the internet of connected vehicles, and routing protocols are presented. Chapter 12 presents various computing models, highlighting cloud, edge, fog and moisture computing and salient features of edge computing. Integration of 5G, IoT and digital communication systems is well explained.

Editors

Dr. Inam Ullah received a bachelor's degree in Electrical Engineering (Telecommunication) from the Department of Electrical Engineering, University of Science and Technology Bannu (USTB), KPK, Pakistan, in 2016 and a master's and a Ph.D. degree in Information and Communication Engineering from the College of Internet of Things (IoT) Engineering, Hohai University (HHU), Changzhou Campus, 213022, China, in 2018 and 2022, respectively. He completed his postdoc with Brain Korea 2021 (BK21) at the Chungbuk Information Technology Education and Research Center, Chungbuk National University, Cheongju 28644, South Korea, from October 2022 to March 31, 2023. He is currently an assistant professor at the Department of Computer Engineering, Gachon University, South Korea. His research interests include Robotics, Internet of Things (IoT), Wireless Sensor Networks (WSNs), Underwater Communication and Localization, Underwater Sensor Networks (USNs), Artificial Intelligence (AI), Big data and Deep learning. He has authored more than 70 peer-reviewed articles on various research topics. He is the reviewer of many prominent journals, including *IEEE Transactions on Industrial Informatics KSII Transactions on Internet & Information Systems, IEEE Transactions on Vehicular Technology, IEEE Transactions on Intelligent Transportation Systems, Transactions on Sustainable Computing, IEEE ACCESS, Sustainable Energy Technologies and Assessments, Future Generation Computer Systems* (FGCS), *Computers and Electrical Engineering* (Elsevier), *Internet of Things (IoT) Journal, Digital Communications & Networks* (Elsevier), *Wireless Communication & Mobile Computing (WCMC), Alexandria Engineering Journal Sensors, Electronics, Remote Sensing, Applied Sciences, Computational Intelligence and Neurosciences*, etc. His awards and honors include the Best Student Award from the University of Science and Technology Bannu (USTB), KPK, Pakistan, in 2015; the Prime Minister Laptop Scheme Award from the University of Science and Technology Bannu (USTB), KPK, Pakistan, in April 2015; Top 10 students award of the College of Internet of Things (IoT) Engineering, Hohai University, China, in June 2019; Top 100 students award of Hohai University (HHU), China, in June 2019; Jiangsu Province Distinguish International Students Award (30,000 RMB) in 2019–2020; Certificate of Recognition from Hohai University (HHU), China, in 2021 and 2022 both; Top 100 students award of Hohai University (HHU), China, in May 2022; Top 10 Outstanding Students Award, Hohai University (HHU), China, in June 2022; and Distinguished Alumni Award from University of Science and Technology Bannu (USTB), KPK, Pakistan, in October 2022.

Dr. Inam Ullah Khan is the founder of the Internet of Flying Vehicles Lab at AI-EYS. Currently, he has been working as Global Mentor/Guest Lecturer at Impact Xcelerator, IE School of Science and Technology, Madrid, Spain. Previously, he was working as Visiting Researcher at King's College London, United Kingdom. Also, he was Faculty Member at different universities in Pakistan, which include the Center for Emerging Sciences, Engineering & Technology (CESET), Islamabad; Abdul Wali Khan University, Garden Campus, Timergara Campus; University of Swat; and Shaheed Zulfikar Ali Bhutto Institute of Science and Technology (SZABIST), Islamabad Campus. He completed his Ph.D. in Electronics Engineering from the Department of Electronics Engineering, Isra University, Islamabad Campus, School of Engineering & Applied Sciences (SEAS). Also, he did his M.S. degree in Electronics Engineering at the Department of Electronics Engineering, Isra University, Islamabad Campus, School of Engineering & Applied Sciences (SEAS). He had a bachelor's degree in Computer Science from Abdul Wali Khan University Mardan, Pakistan. He authored/coauthored more than 50 research articles in reputable journals, conferences and book chapters. His research interests include Network System Security, Intrusion Detection, Intrusion Prevention, Cryptography, Optimization Techniques, WSN, IoT, Mobile Ad Hoc Networks (MANETS), Flying Ad Hoc Networks and Machine Learning. He served in many international conferences as Session Chair/Technical Program Committee Member. Also, he served as Guest Editor with many prestigious international journals. Apart from that, he is General Chair at the International Conference on Trends and Innovations in Smart Technologies (ICTIST'22), virtually from London, United Kingdom. In addition, he also served as Editor of several books. More interestingly, he was invited as a technology expert many times on Pakistan National Television and other media outlets.

Dr. Mariya Ouaissa is currently a professor in Cybersecurity and Networks at the Faculty of Sciences Semlalia, Cadi Ayyad University, Marrakech, Morocco. She earned her Ph.D. degree in Computer Science and Networks in 2019 at the Laboratory of Modelisation of Mathematics and Computer Science from ENSAM-Moulay Ismail University, Meknes, Morocco. She is a networks and telecoms engineer, who graduated in 2013 from the National School of Applied Sciences, Khouribga, Morocco. She is a co-founder and IT consultant at IT Support and Consulting Center. She worked as a visiting professor at the School of Technology of Meknes, Morocco, from 2013 to 2021. She is Member of the International Association of Engineers and the International Association of Online Engineering, and since 2021, she has been an "ACM Professional Member." She is Expert Reviewer with Academic Exchange Information Centre (AEIC) and Brand Ambassador with Bentham Science. She has served and continues to serve on technical program and organizer committees of several conferences and events and has organized many Symposiums/Workshops/Conferences as a general chair and also as a reviewer of numerous international journals. She has made contributions in the fields of information security and privacy, IoT security and wireless and constrained networks security. Her main research topics are IoT, M2M, D2D, WSN, Cellular Networks and Vehicular Networks. She has published over 40 papers (book chapters, international journals and conferences/workshops), 12 edited books and 8 special issues (as guest editor).

Dr. Mariyam Ouaissa is currently an assistant professor in Networks and Systems at ENSA, Chouaib Doukkali University, El Jadida, Morocco. She received her Ph.D. degree in 2019 from the National Graduate School of Arts and Crafts, Meknes, Morocco, and her Engineering degree in 2013 from the National School of Applied Sciences, Khouribga, Morocco. She is a communication and networking researcher and practitioner with industry and academic experience. Her research is multidisciplinary, and it focuses on the IoT, M2M, WSN, vehicular communications and cellular networks, security networks, congestion overload problem and resource allocation management and access control. She is serving as a reviewer for international journals and conferences including *IEEE Access*, *Wireless Communications* and *Mobile Computing*. Since 2020, she has been a member of the "International Association of Engineers (IAENG)" and "International Association of Online Engineering," and since 2021, she has been an "ACM Professional Member." She has published more than 30 research papers (this includes book chapters, peer-reviewed journal articles and peer-reviewed conference manuscripts), 10 edited books and 6 special issues (as guest editor). She has served on Program Committees and Organizing Committees of several conferences and events and has organized many Symposiums/ Workshops/Conferences as a general chair.

Dr. Salma El Hajjami has been an assistant professor and researcher at the Faculty of Science, Ibn Zohr University, Agadir, Morocco, since 2021. She earned her Ph.D. degree in Computer Science in 2021 at the Laboratory of Artificial Intelligence, Data Science and Emerging Systems from ENSA, Sidi Mohammed Ben Abdellah University, Fez, Morocco. She is a computer science engineer, who graduated in 2015 from the National School of Applied Sciences Fez, Morocco. She has previous expertise acting in the Ministry of Interior Morocco as Research and Development Engineer from 2017 to 2021. She is Member of the International Association of Engineers (IAENG) and the International Association of Online Engineering. Dr. Salma has made contributions in the fields of Social Big Data, Semantics Analytics, Anomaly Detection and Imbalanced Big Data published at international conferences and journals. Her main research topics are Machine Learning, Deep Learning, Imbalanced Big Data, Data Science and Blockchain. She has served and continues to serve on technical program and organizer committees of several conferences and also as a reviewer of numerous international journals.

Contributors

Rabia Afzaal
Department of Information Technology
Lahore Garrison University
Lahore, Pakistan

Ijaz Ahamd
Shenzhen College of Advanced Technology
University of Chinese Academy of Sciences
Shenzhen, China

Qasem Abu Al-Haija
Department of Cybersecurity, Faculty of
    Computer & Information Technology
Jordan University of Science and
    Technology
Irbid, Jordan

Wadee Alhalabi
Department of Computer Science,
    Immersive Virtual Reality Research
    Group
King Abdulaziz University
Jeddah, Saudi Arabia

Ahad Alhudali
Department of Computer Science,
    Immersive Virtual Reality Research
    Group
King Abdulaziz University
Jeddah, Saudi Arabia

Abdullah Alwabli
Department of Communication and
    Electronics Engineering, College of
    Engineering in Al-Qunfudhah
Umm Al-Qura University
Mecca, Saudi Arabia

Muhammad Shahid Anwar
Department of AI and Software
Gachon University
Seongnam-si, Republic of Korea

Shahzad Ashraf
NFC Institute of Engineering and
    Technology
Multan, Pakistan

Hania Batool
Department of Computer Science, Faculty
    of ICT
BUITEMS
Quetta, Pakistan

Soukayna Riffi Boualam
Department of Computer Science
Faculty of Sciences
Moulay Ismail University
Meknes, Morocco

Zakaria Boulouard
LIM
Hassan II University
Casablanca, Morocco

Ahyoung Choi
Department of AI and Software
Gachon University
Seongnam-si, Republic of Korea

Sarah El Himer
Department of Electrical Engineering
Faculty of Sciences and Technologies
Sidi Mohamed Ben Abdellah University
Fez, Morocco

Hina Hassan
College of Life Science and Technology
Harbin Normal University
Harbin, China

Sadaf Hussain
Department of Computer Science
Lahore Garrison University
Lahore, Pakistan

Waleed M. Ismael
Department of Information Technology,
    Faculty of Engineering
Azal University for Human Development
Sanaa, Yemen

Adnan Khalil
Department of Computer Science and
    Information Technology
University of Malakand
Khyber Pakhtunkhwa, Pakistan

Irshad Khalil
Department of Health Sciences and
    Technology, Gachon Advanced Institute
    for Health Sciences and Technology
    (GAIHST)
Gachon University
Incheon , Republic of Korea

Habib Khan
Department of Computer Science
Islamia College University
Peshawar, Pakistan

Ijaz Khan
School of Electronics and Information
    Engineering
Harbin Institute of Technology
Harbin, China

Inam Ullah Khan
Department of Computer Science
Szabist University
Islamabad, Pakistan

Mosa Khan
Department of Computer Science and
    Information Technology
University of Malakand
Khyber Pakhtunkhwa, Pakistan

Muhammad Adnan Khan
Riphah School of Computing &
    Innovation, Faculty of Computing
Riphah International University
Lahore, Pakistan
and
School of Computing
Skyline University College
Sharjah, United Arab Emirates
and
Department of Software, Faculty of
    Artificial Intelligence and Software
Gachon University
Seongnam, Republic of Korea

Rahim Khan
College of Information and
    Communication Engineering
Harbin Engineering University
Harbin, China

Xuefei Ma
College of Information and
    Communication Engineering
Harbin Engineering University
Harbin, China

Yong-Kui Ma
School of Electronics and Information
    Engineering
Harbin Institute of Technology
Harbin, China

Adila Mehdi
Department of Computer Science, Faculty
    of ICT
BUITEMS
Quetta, Pakistan

Muhammad Muneer
Department of Statistics & Data Science
University of Mianwali
Mianwali, Pakistan

Mariya Ouaissa
Computer Systems Engineering Laboratory
Cadi Ayyad University
Marrakech, Morocco

Mariyam Ouaissa
Laboratory of Information Technologies
Chouaib Doukkali University
El Jadida, Morocco

Zahid Rasheed
School of Electronics and Information
    Engineering
Harbin Institute of Technology
Harbin, China

Faisal Rehman
Department of Statistics & Data Science
University of Mianwali
Mianwali, Pakistan
and
Department of Robotics & Artificial
    Intelligence
National University of Science and
    Technology (NUST)
Islamabad, Pakistan

Naveed Riaz
Department of Mechanical Engineering
National University of Science and
    Technology (NUST)
Islamabad, Pakistan

Amal Saif
Department of Computer Science, King
    Hussein School of Computing Sciences
Prince Sumaya University for Technology
Amman, Jordan

Ahthasham Sajid
Department of Software Engineering,
    Faculty of Computing
Capital University of Science and
    Technology
Islamabad, Pakistan

Muhammad Hamza Sajjad
Department of Statistics & Data Science
University of Mianwali
Mianwali, Pakistan

Wasswa Shafik
School of Digital Science
Universiti Brunei Darussalam
Bandar Seri Begawan, Brunei
and
Dig Connectivity Research Laboratory
    (DCRLab)
Kampala, Uganda

Muhammad Shafiq
Cyberspace Institute of Advanced
    Technology
Guangzhou University
Guangzhou, China

Tanweer Sohail
Department of Mathematics
University of Jhang
Jhang, Pakistan

Sher Taj
Software College
Northeastern University
Shenyang, China
and
Daqing Normal University
Harbin, China

Yuning Tao
School of Electric Power
South China University of Technology
Guangzhou, China

Habib Ullah
College of Electronics and Information
    Engineering
Nanjing University of Aeronautics and
    Astronautics
Nanjing, China

Inam Ullah
Department of Computer Engineering
Gachon University
Seongnam, Republic of Korea

Fazal Wahab
College of Computer Science and
    Technology
Northeastern University
Shenyang, China

I

AIoT and Challenges for Future Communication Systems

Artificial Intelligence in the Internet of Things, Recent Challenges and Future Prospects

Fazal Wahab
Northeastern University

Mosa Khan
University of Malakand

Inam Ullah
Gachon University

Yuning Tao
South China University of Technology

1.1 INTRODUCTION

In this chapter, we will explain Artificial Intelligence (AI) in the Internet of Things (IoT), Machine Learning (ML), and their applications in IoT.

1.1.1 Artificial Intelligence in the Internet of Things

The IoT and AI are two significant developments in the ever-evolving field of technology that are having a profound impact on the world as we know it. The IoT is no longer a futuristic theory but a pervasive reality that has profoundly altered the way we interact with the physical world. The integration of a wide range of devices, sensors, and systems creates a complex network that permeates our everyday lives. This network facilitates the gathering,

DOI: 10.1201/9781032648309-2

3

dissemination, and independent analysis of data on an unparalleled level. The IoT has significantly transformed the way individuals establish connections, engage in communication, and interact with the tangible environment [1]. The acronym IoT refers to a sophisticated network of interconnected objects, sensors, and systems with autonomous capabilities to collect, exchange, and analyse data. In contrast, artificial intelligence (AI) comprises a variety of technologies, including ML and deep learning, which facilitate the emulation of human intellect by machines. These technologies enable machines to acquire knowledge from data and utilise it to make informed judgements. The integration of AI and IoT, referred to as AIoT, offers substantial prospects for enhancing efficacy, security, and convenience across several industries, such as manufacturing, healthcare, and intelligent communities.

The immense potential of AIoT becomes apparent when we contemplate the boundless opportunities it holds for the future. The potential of AIoT capabilities is set to be enhanced by the emergence of technologies such as 5G connectivity and edge computing, hence expanding the boundaries of what can be achieved. AI algorithm sophistication continues to rise, promising deeper insights from IoT-generated data, more exact forecasts, and a higher level of automation. The stage is set for more innovation, with AIoT serving as the driving force behind transformative advances in various industries and aspects of human existence [2]. Together, AI and IoT build a new paradigm in which data play a central role, intelligence permeates all aspects, and the only thing limiting the possibilities is our ingenuity. In the dynamic environment of technological advancements, the partnership between human creativity and machine intelligence is limitless, leading us to a future where the integration of AI and IoT fundamentally transforms our global reality. Every industry and area of study feels the effects of smarter applications' improved use of data insights. Technology and methods for learning from and acting on the massive volumes of data created daily are the main forces behind this shift. Deep neural networks, conventional ML methods, and scalable GPU computing have all contributed to the tremendous development and reduced impediments in adoption [3]. Python is widely regarded as the most prevalent programming language for scientific computing, data research, and ML due to its extensive collection of low-level libraries and intuitive high-level application programming interfaces (APIs).

This study thoroughly examines the interconnected domains of AI and IoT. This analysis will explore the complex dynamics of their interaction, highlighting the significant influence they have already exerted across multiple industries and providing information on future prospects. The exploration will encompass the domains of enhanced data analytics, energy optimisation, healthcare, smart cities, and manufacturing, allowing us to observe the integration of intelligence via AIoT into the fundamental structure of these areas. However, it is essential to acknowledge that along with the potential benefits, significant responsibilities must be addressed. As we embrace this technological revolution, we will carefully discuss the various problems and factors that come with it. These include ensuring data interoperability, security, scalability, and energy efficiency.

1.1.2 Artificial Intelligence

The discipline within computer science, commonly referred to as AI, is dedicated to the development of computer programs and hardware capable of executing activities that have traditionally been associated with human intelligence and behaviour. In the middle of the twentieth century, researchers developed a subfield of AI called ML, which sought to mimic the brain's conceptual structure and function in order to create AI [4–6]. The utilisation of ML continues to be imperative in the advancement of AI. ML is widely recognised as a scientific discipline that centres on the advancement of computer models and algorithms capable of executing specific tasks, often including the identification of patterns, without the requirement of explicitly coding them. Computer programming is a fascinating and complicated field focusing on automating and improving routine tasks. For the purpose of automating mail sorting, for instance, a programmer can employ zip code recognition software. Developing a comprehensive set of rules that can be implemented in a computer program to effectively execute this activity is frequently a laborious and demanding process. In the present context, the term "ML" pertains to the research and development of technologies that enable computers to automatically make complex decisions by recognising patterns in labelled data and drawing inferences about the relevance of those patterns without being explicitly programmed with the rules to follow. In the previous example of zip code identification, ML was used to develop a model from labelled cases, resulting in very accurate recognition of both machine-generated and hand-written zip codes.

The term "artificial intelligence" (AI) refers to a type of technology that aspires to endow computers with the ability to reason in a manner like that of humans. The process by which different industries are digitally reforming themselves will be sped up as a result of this recent revelation. Enabling interconnectedness and augmenting decision-making capabilities across various entities, including individuals, animals, plants, robots, appliances, natural elements, and infrastructure, have the potential to transform the global landscape into a more self-sustaining ecosystem [7–9]. This transformation will take place regardless of whether it is people, animals, plants, machines, or appliances. Suppose we want to achieve our goal of providing the environment and the physical things in its full autonomy. In that case, the system will need to incorporate not only a data analysis (DA) module but also an ML module that models human learning. ML is credited for developing methodologies that facilitate autonomous and self-sustaining learning in different components and devices within a network. On the contrary, data analytics (DA) focuses on evaluating and examining the data generated over a period of time to identify historical patterns and enhance future efficiency and effectiveness. The observed trend has exhibited growth, and ongoing efforts are being made to integrate ML and DA into the sensors and embedded systems of intelligent systems. The scientific underpinnings of AI are highly intriguing, prompting a reevaluation of our understanding of existential concepts, such as the purpose of human existence and the nature of employment [10]. The lightning-fast speed at which ML and DA are driving AI makes it necessary for us to have a dialogue about the tendencies, challenges, and dangers that will eventually become more serious.

Despite the efficient elimination of redundant human labour and the ability of AI-based systems to provide outputs in a shorter timeframe, it can be argued that human inventiveness will always serve a distinct purpose in productive endeavours. The vast bulk of ongoing research and development in AI fall under the umbrella term of "Narrow AI." This suggests that the employment of technology is only beneficial to a select few endeavours and not all of them. Despite this, our goal is to achieve something that is orders of magnitude more significant than that. Keeping this in view, experts from a wide variety of fields have worked together to accelerate the development of AI. The multidisciplinary nature of AI has been reinforced via the collaborative endeavours of numerous academic fields, such as statistics, philosophy, physics, computer science, sociology, mathematics, biology, and psychology, among various others. The accumulation of data in each of these areas is crucial in the growth of intelligence [11,12]. To understand what the underlying principles are, some sort of analysis of these facts must be performed. In spite of the fact that the human brain is fully capable of performing the task, it takes a significant amount of time to do so.

This phenomenon can be ascribed to the existence of numerous undesirable attributes in the data acquired from sources in the real world, encompassing but not restricted to the subsequent:

- Unstructured in its nature

- Huge in its volume

- A wide variety of data sources

- A requirement for real-time processing

- Ongoing and consistent change

Furthermore, there exist supplementary attributes, such as volatility and virility, among others. AI is a systematic approach that aims to optimise the utilisation of data in a manner that ensures comprehensibility for data providers, allows for modifications to rectify errors, has value within the specific application environment, and possesses meaningfulness. Hence, AI exhibits a significant reliance on the approaches employed within the field of data science. In a more expansive context, data science can be delineated as the scientific discipline concerned with the development of tools and methodologies for the purpose of scrutinising vast quantities of data with the objective of extracting valuable insights. Consequently, the discipline represents an amalgamation of various distinct research subdisciplines. The primary source of inspiration for the advancement of tools stems from the domain of computer science, which predominantly focuses on the efficacy of algorithms and the expandability of storage [13].

1.1.3 Machine Learning

ML is a burgeoning discipline within the realm of computer science that facilitates the capacity of inanimate computer systems to gain the aptitude for learning all without necessitating the explicit implementation of code. Digital computers have transformed practically all areas of economic activity during the past few decades [8,14,15]. We are currently on the verge of a significant and accelerated transformation due to recent advancements in ML, which have the potential to expedite the process of automation. The absence of consensus regarding the domains in which ML systems excel has resulted in a corresponding lack of consensus regarding the specific anticipated impacts on labour and the broader economy. Although

ML is often categorised as a "general purpose technology" akin to the steam engine and electricity, facilitating a wide range of innovative developments and prospects, this statement remains valid. The consequences pertaining to employment exhibit a level of complexity that surpasses the simplistic focus on replacement and substitution. On the contrary, it is possible that specific individuals may lack these particular attributes. Although the notion of the "end of work" is not expected to occur in the near future, it is important to recognise the significant long-term implications of ML on the economy and the labour force.

When addressing the potential benefits and drawbacks of ML and its possible effects on the economy, there are two fundamental considerations to keep in mind. Significant progress is yet to be made in developing highly intelligent machines [16]. According to scholarly research, it has been observed that humans possess a more comprehensive array of skills compared to machines [17]. The impact of technological advancements on wage inequality is significant despite the positive effects they have had on income and living standards. The initial wave of pre-ML information technology (IT) systems, mainly, has generated trillions of dollars in economic value. However, there is substantial evidence supporting the argument that technological progress has played a role in exacerbating wage disparities. Numerous reasons, including the expanding process of globalisation, contribute to the presence of inequality. However, it is important to acknowledge that the economic consequences of this phenomenon can be highly disruptive. The rapid and profound changes that can be brought about by ML are primarily responsible for this upheaval. These changes can take place in as little as ten years. Policymakers, business executives, engineers, and academics all have a role to play here.

As robots replace more and more human workers in a process or industry, the value of non-SML occupations may rise. Technology can help people reach their full potential and introduce them to new ways of doing things. Therefore, even within highly mechanised professions, the overall effect on the need for labour could be either positive or negative. Jobs that are close substitutes for the capabilities of ML are more likely to see a drop in labour demand, whereas jobs that are complements to these systems are more likely to experience an increase in demand, but the broader economic effects can be complex [14,18,19]. When an ML system can complete a task as well as a human but for less money, profit maximisation is the goal. More and more bosses and CEOs are looking for ways to automate human jobs. Impacts on productivity, prices, labour demand, and the structure of diverse industries could have far-reaching effects on the economy.

In comparison to traditional computer applications, the IoT introduces a situation where the speed, diversity, volume, and intricacy of data are so immense that it becomes unfeasible for a human programmer to provide specific, detailed, and precise job requirements that can be utilised to execute the tasks. Consequently, the concept of ML is around the acquisition of implicit learning abilities, enabling a computer or system to autonomously educate itself, adapt to its surroundings, and make independent decisions. The smart concept in CPS or IoT can thus be made up for by using ML in this way [13,20].

ML is an approach to attaining AI that revolves around the concept of providing computers with access to data, enabling them to acquire knowledge autonomously. This concept is known as the "data-driven learning hypothesis." Researchers often assume a premise that they possess the knowledge necessary to predict the future development of AI that can rival human capabilities. Undoubtedly, significant advancements are being made towards the objective at an increasingly rapid pace. A significant proportion of the advancements achieved in recent years can be ascribed to the fundamental paradigm shifts that have occurred in our comprehension of the operational mechanisms of AI. ML has predominantly been the catalyst for these advances. Consequently, the decision to associate ML with the capacity to bestow intelligence upon computers is a justifiable one.

Even just a few decades ago, the idea of being able to have a video conversation with family members who lived on another continent was unfathomable to everyone. These days, it is an everyday occurrence. All of these things are the result of technology becoming more affordable and new devices appearing on the market with enhanced and augmented capabilities [21]. Paying bills, sending emails, transferring money, or even scheduling a taxi are some things that can be accomplished with a button on a smartphone. The term "Internet of Computers" (IoC) has been in use since 1991, and it steadily expanded in scope as an ever-increasing number of individuals began to make use of it.

1.1.4 Reinforcement Learning: A Type of ML

Reinforcement learning, a captivating domain within ML, exhibits notable parallels with the learning processes observed in humans. Several research presented the reinforcement learning method and effectively implemented it in the context of checkers, wherein a linear value function was employed to guide decision-making [22,23]. The field of ML includes reinforcement learning. Reinforcement learning differs from other learning approaches because it evaluates new actions based on previous ones. Taking action is required because it is the correct choice. Since the agent is clueless, they should not try anything. Instead, it investigates how to maximise positive outcomes through specific forms of action. Therefore, RL is a system of trial-and-error learning where experimentation and response are the way to go.

1.2 INTERNET OF THINGS

The IoT is a transformative revolution in which interconnected physical gadgets, buildings, vehicles, and diverse objects form a networked system. These entities are outfitted with software, sensors, and networking capabilities, facilitating their ability to communicate and share data. These technologically advanced devices are specifically engineered to gather and share data through internet connectivity, facilitating their ability to communicate with one another and central systems [24–26]. The establishment of connectivity enables the ability to monitor, control, and automate many operations in real time, resulting in enhanced efficiency, convenience, and insights across multiple domains. The IoT has become pervasive in both our personal lives and various industries. It encompasses a range of applications, such as smart home devices that adapt lighting and temperature settings according to user preferences, as well as industrial sensors that enhance manufacturing processes. This technology holds the potential to create a future where the boundaries between the physical and digital realms seamlessly merge, resulting in advancements that enhance our overall well-being and revolutionise business operations.

Data science and AI studies have been concentrating on this problem as their primary focus. Therefore, IoT and AI together may represent a significant step forward. It's not just about cutting costs, being resourceful, reducing manual labour, or following the latest fashion. More than just making people's lives simpler is at stake here. Many serious issues, such as ethical and security problems, beset the IoT and are not going away anytime soon. What counts is not how fascinating the IoT plus AI appears to be but rather how the ordinary person perceives it [27]. As time goes on, the internet's capabilities will shift from those associated with the "Internet of Computers (IoC)" to those associated with the "IoT." Furthermore, integrating many elements, such as infrastructure, embedded devices, human agents, intelligent objects, and physical surroundings, leads to the emergence of highly interconnected systems commonly referred to as Cyber-Physical Systems (CPS). In the future, we will live in a world where the "Internet of Everything" is seamlessly connected to the "Smart Cyber-Physical Earth." There is hope that "data science," in concert with IoT and Cyber-Physical Systems, will spark the next "smart revolution." The primary concern arises in determining the most effective approach to manage the vast quantities of data given the limited processing capabilities presently accessible.

Being "smart" as a notion fascinates us much. However, the current state of our resources is still quite a way off from matching human intelligence. Let's use a smartphone as an example, which, despite its "smart" label, can't handle anything independently. For instance, the gadget cannot put itself into "silent mode" for notifications and message alerts when the owner is behind the wheel of a car. It would be a more useful piece of technology if it could reduce the number of disruptions caused by notifications while the owner is behind the wheel. For this system to operate, it is necessary to establish a wireless connection between the individual, their smartphone, and the vehicle. If the smartphone's owner suddenly gets ill, a distress call should be placed to a member of the owner's family or a nearby medical facility.

To actualise this vision, it will once again be imperative to establish specific connections and acquire comprehensive data pertaining to the individuals comprising the family unit and the healthcare facilities involved. By consistently presenting such instances, it becomes evident that in order to satisfy various sets of criteria, practically all entities in the physical realm must establish connections with one another. If we want these things to have intelligence, we are going to have to resort to some form of AI.

1.2.1 Architecture of IoT

For the IoT, there is no generally accepted blueprint for the network's infrastructure. Researchers have come up with a wide variety of innovative architectural concepts.

1.2.2 Three- and Five-Layer Architectures

As illustrated in Figure 1.1, the fundamental arrangement consists of three distinct layers [3–6]. The origins of this subject can be traced back to the initial phases of scientific inquiry. The system is structured into three distinct layers, namely, the "perception" layer, the "network" layer, and the "application" layer. The physical layer, also known as the perception layer, encompasses a set of sensors responsible for detecting and gathering data about

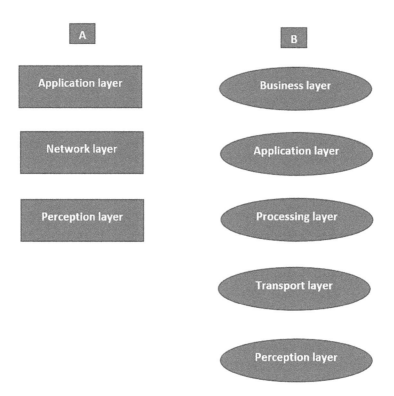

FIGURE 1.1 Different IoT architectures.

the immediate environment. The device has the capability to see and analyse environmental factors, as well as identify and interact with other entities possessing intelligence [28,29]. The network layer is responsible for managing connectivity to various intelligent creatures, network nodes, and servers. The utilisation of its capabilities extends to the transmission and analysis of sensor data as well. The application layer (Layer III) offers program-specific functions to the user. The IoT comprises a diverse array of applications, such as the implementation of intelligent agricultural practices, the development of technologically advanced urban environments, the integration of technology in educational environments, the creation of intelligent residential spaces, and the establishment of advanced healthcare systems. Figure 1.1 represents the three- and five-level IoT architectures.

1.2.3 How Does IoT Work?

Each IoT echo system has a slightly distinct method of operation. However, their fundamental principles of operation are the same. Devices, such as smartphones, digital watches, and electronic appliances, begin the IoT's workflow by securely exchanging data with the IoT platform. To ensure the provision of highly pertinent information to devices, the platforms employ data aggregation techniques that encompass a diverse array of sources [30]. Figure 1.2 illustrates the working environment of IoT.

The inception of the IoT can be traced back to the development of portable telecommunication devices and other interconnected devices. Throughout its development, the reach of this phenomenon has broadened due to the increased affordability and accessibility of

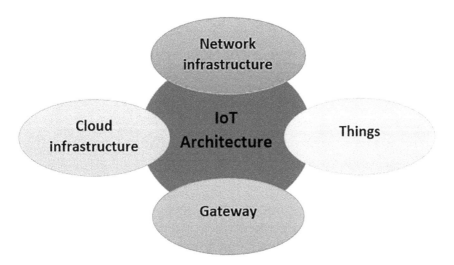

FIGURE 1.2   IoT working environment.

personal computers, mobile phones, laptops, and tablet computers to the general population. Based on the forecasts provided by Gartner Inc., the global use of connected devices
is expected to reach 6.4 billion in 2016, reflecting a growth rate of 30% compared to the
previous year. According to the cited source [31], the figure is projected to increase to 20.8
billion by 2020. In 2016, it was observed that daily, an average of more than 5.5 million new
devices were being connected to the internet. These data highlight the significant capacity and possibilities presented by the IoT. The IoT encompasses diverse sectors due to its
fundamental characteristic of facilitating continuous connectivity among various things.
Consequently, the IoT might be construed as a means of amalgamating several disciplines.
Figure 1.1 provides a list of domains that are illustrative of those that are included in the
IoT [32]. These domains make up the IoT. Most of these fields have similar conceptual
underpinnings and methodological techniques. The IoT is basically a network that connects humans and a variety of inanimate and live creatures, such as appliances, crop fields,
plants, and animals. It is also known as the Internet of Everything (IoE). Using intelligent
devices hooked to both technologies and able to transmit, receive, and process data is the
means by which humans are connected to them. These intelligent things can send, receive,
and process data. These intelligent objects represent the entity (either a living being or an
inanimate object) to which they are connected in the network. This entity may be a living
being or an inanimate object [33,34].

Over the past decade, there has been a notable surge in the quantity of internet-connected
gadgets, coinciding with the proliferation of concerns pertaining to cybersecurity. AI plays
a pivotal role in the vanguard of cybersecurity endeavours due to its capability to facilitate
the creation of intricate algorithms aimed at safeguarding critical infrastructure, including
the IoT. Nevertheless, hackers have successfully harnessed the potential of AI and have been
employing adversarial AI in their illicit activities. The utilisation of AI by cyber criminals
has been observed. This review study integrates information from many prior surveys and
research articles about IoT, AI, and attacks involving AI, as referenced by source [35–37].

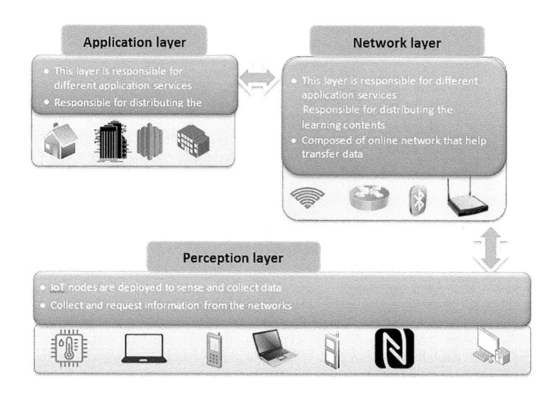

FIGURE 1.3   IoT architecture.

This was done with the intention of providing a comprehensive presentation and synthesis of relevant work on these topics. Figure 1.3 illustrates the IoT's basic architecture.

Since its inception in the early 2010s, the IoT has been rapidly spreading and is already widespread in many residences and places of business. The IoT can be defined as "a system in which objects embedded with network connectivity and individually generated identification numbers (UIDs) exchange data with one another without the need for human intervention" [38,30]. This definition takes into consideration the dynamic nature of the IoT, acknowledging its transformation and development since its first conceptualisation. Frequently, this phenomenon manifests as an interaction between a human entity and a central device or software application, commonly found in the form of a smartphone application. Subsequently, this connection facilitates the transmission of data and commands to one or several edge IoT devices [39]. Peripheral devices possess the capability to execute essential functions and transfer data to the central device or application, enabling the user to assess afterwards and analyse the information.

The idea of the IoT makes it possible to improve accessibility, integrity, availability, scalability, confidentiality, and interoperability among linked devices. However, since they are still relatively new and not enough time has been spent defining security standards and processes, IoT is vulnerable to being attacked. Cybercriminals have the potential to employ a diverse range of cyberattacks against IoT devices, contingent upon the specific component of the system they aim to exploit and the objectives they seek to accomplish [40–43]. As a result, a significant amount of research has been conducted to tackle the matter of augmenting the security of the IoT. This section encompasses a range of methodologies that employ AI for the purpose of safeguarding the IoT infrastructure against malevolent hackers. In contrast, the IoT provides hackers with a significant advantage as they only need to identify a single vulnerability to target an entire network, while cybersecurity experts are tasked with safeguarding several entities. Consequently, there has been a rise in the utilisation of AI by malevolent actors in the realm of cybersecurity as they endeavour to evade detection by intricate algorithms designed to identify abnormal activities [38]. They do this to circumvent the algorithms.

1.3 AIoT IN DIFFERENT SECTORS

AIoT, the combination of AI and IoT, is a revolutionary trend that is changing many different sectors and aspects of people's daily lives.

- **The Utilisation of Advanced Data Analytics to Gain Deeper and More Comprehensive Insights:** One of the most notable features of AI in the IoT is its ability to derive essential insights from the vast quantities of data produced by IoT devices [44,45]. AI algorithms possess the capability to evaluate the aforementioned data in order to discern patterns, irregularities, and connections. This ability facilitates the implementation of predictive maintenance, resource optimisation, and instantaneous decision-making.

- **Energy Efficiency and Optimisation:** The integration of AI and IoT has brought about a significant transformation in the field of energy management systems. The utilisation of AI-driven analysis in examining energy consumption patterns enables the implementation of immediate adjustments, hence mitigating inefficiencies and decreasing expenses in smart grids and buildings.

- **Transforming Medical Practices:** The utilisation of IoT devices, such as wearable technology and remote monitoring systems, is significantly transforming the landscape of healthcare delivery [46,47]. The utilisation of AI in the analysis of patient data has facilitated the timely identification of diseases, the development of tailored treatment strategies, and the enhancement of patient outcomes.

- **Intelligent Communities Progress:** The advancement of smart cities is significantly influenced by the integration of AIoT. IoT sensors are utilised to gather data pertaining to traffic patterns, pollution levels, and energy usage. Simultaneously, AI algorithms are employed to optimise traffic flow, mitigate emissions, and improve urban planning [48–50].

- **Industry 4.0:** The integration of AIoT is driving the transformation of the manufacturing sector, marking the advent of the fourth industrial revolution, which is also known as Industry 4.0. The implementation of this technology enables the optimisation of predictive maintenance, quality control, and autonomous manufacturing processes, resulting in enhanced productivity and minimised periods of operational inactivity [51–54].

1.4 CHALLENGES AND CONSIDERATIONS

The potential of AI in the IoT is vast, yet various problems and considerations accompany its implementation.

- **Data Protection and Privacy:** The proliferation of data created by IoT devices gives rise to substantial apprehensions over preserving data security and privacy. The implementation of robust procedures is crucial in securing sensitive information.

- **Scalability:** The problem of controlling the growing number of networked devices and guaranteeing smooth integration with AI systems becomes significant as IoT networks expand in size and complexity.

- **Interoperability:** Interoperability is a critical factor in maximising the capabilities of varied IoT devices and AI platforms. The process entails creating effective communication and collaboration among various technologies.

- **Energy Efficiency:** The consideration of energy efficiency is of utmost importance for IoT devices that are dependent on battery power. In order to optimise longevity and reduce the need for maintenance, it is crucial to utilise energy-efficient AI algorithms.

1.5 FUTURE PROSPECTS

The symbiotic relationship between AI and IoT is developing rapidly. The possibilities of AIoT will be further enhanced by emerging technologies such as 5G, 6G connectivity, and edge computing. With the increasing sophistication of AI algorithms, there is a potential for unlocking profound insights from IoT data. This has the potential to enhance the accuracy of forecasts and improve automated processes, hence fostering innovation across diverse industries.

1.6 SUMMARY

The integration of AI with the IoT, known as AIoT, is a transformative occurrence that is bringing about significant changes in several industries and impacting our daily lives. The integration of AI with the IoT (AIoT) holds the potential to bring about significant transformations in multiple sectors, including data analytics, energy efficiency, healthcare, and urban development. Nonetheless, it also poses certain difficulties pertaining to the protection of data privacy, the ability to handle larger volumes of data, the compatibility between different systems, and the energy efficiency of the technology. The ongoing development of AI and IoT technologies holds significant promise for innovation and beneficial societal outcomes, therefore establishing AIoT as a crucial area of academic inquiry and practical implementation. This article comprehensively analyses the intricate interplay between AI and IoT, delving into their collective potential, existing implementations, accompanying obstacles, and prospective developments. This article seeks to provide insights into AI's dynamic evolution and crucial function within the IoT domain. By comprehensively evaluating relevant literature, this chapter attempts to highlight the significance of AI in the context of linked smart systems.

REFERENCES

[1] Raschka, S, Patterson J, Nolet C. Machine learning in Python: Main developments and technology trends in data science, machine learning, and artificial intelligence. *Information.* 2020;11(4):193.

[2] McCulloch WS, Pitts W. A logical calculus of the ideas immanent in nervous activity. *The Bulletin of the Mathematical Biophysics.* 1943;5:115–133.

[3] Rosenblatt F. The perceptron: A probabilistic model for information storage and organization in the brain. *Psychological Review.* 1958;65:386.

[4] LeCun Y, Boser B, Denker JS, Henderson D, Howard RE, Hubbard W, Jackel LD. Backpropagation applied to handwritten zip code recognition. *Neural Computation.* 1989;1:541–551.

[5] Brynjolfsson E, Mitchell T. What can machine learning do? Workforce implications. *Science.* 2017;358(6370):1530–1534.

[6] Wahab F, Ullah I, Shah A, Khan RA, Choi A, Anwar MS. Design and implementation of a real-time object detection system based on a single shot detector and OpenCV. *Frontiers in Psychology.* 2022;13:1039645.

[7] Khan HU, Hussain A, Nazir S, Ali F, Khan MZ, Ullah I. A service-efficient proxy mobile IPv6 extension for IoT domain. *Information.* 2023;14(8):459.

[8] Mazhar T, Talpur DB, Shloul TA, Ghadi YY, Haq I, Ullah I, Ouahada K, Hamam H. Analysis of IoT security challenges and its solutions using artificial intelligence. *Brain Sciences.* 2023;13(4):683.

[9] Khan I, Tian YB, Ullah I, Kamal MM, Ullah H, Khan A. Designing of E-shaped microstrip antenna using artificial neural network. *International Journal of Computing, Communication and Instrumentation Engineering.* 2018;5(1):23–26.

[10] Tufail AB, Ma YK, Kaabar MK, Martínez F, Junejo AR, Ullah I, Khan R. Deep learning in cancer diagnosis and prognosis prediction: A minireview on challenges, recent trends, and future directions. *Computational and Mathematical Methods in Medicine.* 2021;2021:28 pages.

[11] Legg S, Hutter M, A collection of definitions of intelligence. *Frontiers in Artificial Intelligence and Applications.* 2007;157:17.

[12] Wahab F, Zhao Y, Javeed D, Al-Adhaileh MH, Almaaytah SA, Khan W, Saeed MS, Kumar Shah R. An AI-driven hybrid framework for intrusion detection in IoT-enabled E-health. *Computational Intelligence and Neuroscience.* 2022;2022:11 pages.

[13] Ng A. What artificial intelligence can and can't do right now. *Harvard Business Review.* 2016;9:1–4.

[14] Jia, J, Wang W. Review of reinforcement learning research. In *2020 35th Youth Academic Annual Conference of Chinese Association of Automation (YAC),* Zhanjiang, China, 2020, pp. 186–191. IEEE.

[15] Liu X, Zhao Y, Xu T, Wahab F, Sun Y, Chen C. Efficient false positive control algorithms in big data mining. *Applied Sciences.* 2023;13(8):5006.

[16] Khan I, Tian YB, Vllah H, Rahman SU, Kamal MM. Design an annular ring microstrip antenna based on an artificial neural network. In *2018 2nd IEEE Advanced Information Management, Communicates, Electronic and Automation Control Conference (IMCEC),* Xi'an, China, May 25, 2018, pp. 2033–2037. IEEE.

[17] Khan S, Ullah I, Ali F, Shafiq M, Ghadi YY, Kim T. Deep learning-based marine big data fusion for ocean environment monitoring: Towards shape optimization and salient objects detection. *Frontiers in Marine Science.* 2023;9:1094915.

[18] Abideen ZU, Mazhar T, Razzaq A, Haq I, Ullah I, Alasmary H, Mohamed HG. Analysis of enrollment criteria in secondary schools using machine learning and data mining approach. *Electronics.* 2023;12(3):694.

[19] Ghosh A, Chakraborty D, Law A. Artificial intelligence in Internet of things. *CAAI Transactions on Intelligence Technology*. 2018;3(4):208–218.

[20] Michalski RS, Carbonell JG, Mitchell TM. *Machine Learning: An Artificial Intelligence Approach (Volume I)* (Vol. 1). Elsevier, 2014.

[21] Theodoridis S, Koutroumbas K. *Pattern Recognition*. Elsevier Science, Amsterdam, The Netherlands, 2008.

[22] Li L, Zhao Y, Li Y, Wahab F, Wang Z. The most active community search in large temporal graphs. *Knowledge-Based Systems*. 2022;250:109101. Elsevier, Amsterdam, The Netherlands.

[23] Huang M, Zhao Y, Wang Y, Wahab F, Sun Y, Chen C. Multi-graph multi-label learning with novel and missing labels. *Knowledge-Based Systems*. 2023;276:110753.

[24] Hastie T, Tibshirani R, Friedman J. *The Elements of Statistical Learning: Data Mining, Inference, and Prediction*. Springer Series in Statistics, Springer, New York, 2013.

[25] Holler J, Tsiatsis V, Mulligan C, Avesand S, Karnouskos S, Boyle D. *From Machine-to-Machine to the Internet of Things: Introduction to a New Age of Intelligence*. Academic Press, Cambridge, 2014.

[26] Yu C, Liu H, Wahab F, Ling Z, Ren T, Ma H, Zhao Y. Global triangle estimation based on first edge sampling in large graph streams. *The Journal of Supercomputing*. 2023;79:14079–14116.

[27] Kaplan J. *Artificial Intelligence: What Everyone Needs to Know. What Everyone Needs To Know*. Oxford University Press, Oxford, 2016.

[28] Wahab F, Khan I, Hussain T, Amir A. An investigation of cyber attack impact on consumers' intention to purchase online. *Decision Analytics Journal*. 2023;8:100297.

[29] Jha S, Seshia SA. A theory of formal synthesis via inductive learning. *Acta Informatica*. 2017; 54(7):693–726.

[30] Gupta D, Juneja S, Nauman A, Hamid Y, Ullah I, Kim T, Tag eldin EM, Ghamry NA. Energy saving implementation in hydraulic press using industrial Internet of Things (IIoT). *Electronics*. 2022;11(23):4061.

[31] Kuzlu M, Fair C, Guler O. Role of artificial intelligence in the Internet of Things (IoT) cyber-security. *Discover Internet of Things*. 2021;1:1–14.

[32] Shah A, Ali B, Wahab F, Ullah I, Amesho KT, Shafiq M. Entropy-based grid approach for handling outliers: A case study to environmental monitoring data. *Environmental Science and Pollution Research*. 2023;30:1–20.

[33] Lu Y, et al. A survey of emerging 5G technologies. *IEEE Journal on Selected Areas in Communications*. 2017;123:102917.

[34] Hussain T, Yang B, Rahman HU, Iqbal A, Ali F, Shah B. Improving source location privacy in social Internet of Things using a hybrid phantom routing technique. *Computers & Security* 2022;123:102917.

[35] Evans D. *The Internet of Things: How the Next Evolution of the Internet is Changing Everything*. Cisco Internet Business Solutions Group, Cisco, California, 2011.

[36] Rouse M. What is IoT (Internet of Things) and how does it work? IoT agenda, TechTarget. https://www.internetofthingsagenda.techtarget.com/defnition/Internet-of-Things-IoT. Accessed 11 February 2020.

[37] Yin Y, Shi Y, Zhao Y, Wahab F. Multi-graph learning-based software defect location. *Journal of Software: Evolution and Process*. 2023:e2552.

[38] Khan WU, Imtiaz N, Ullah I. Joint optimization of NOMA-enabled backscatter communications for beyond 5G IoT networks. *Internet Technology Letters*. 2021;4(2):e265.

[39] Mazhar T, Irfan HM, Khan S, Haq I, Ullah I, Iqbal M, Hamam H. Analysis of cyber security attacks and its solutions for the smart grid using machine learning and blockchain methods. *Future Internet*. 2023;15(2):83.

[40] Ullah I, Qian S, Deng Z, Lee JH. Extended Kalman filter-based localization algorithm by edge computing in wireless sensor networks. *Digital Communications and Networks*. 2021;7(2):187–195.

[41] Linthicum D. App nirvana: When the internet of things meets the API economy. https://techbeacon.com/app-dev-testing/app-nirvana-wheninternet-things-meets-api-economy. Accessed 15 November 2019.

[42] Lu Y, Xu LD. Internet of Things (IoT) cybersecurity research: A review of current research topics. *IEEE Internet of Things Journal.* 2019;6(2):2103–2115.

[43] Vorakulpipat C, Rattanalerdnusorn E, Thaenkaew P, Hai HD. Recent challenges, trends, and concerns related to IoT security: An evolutionary study. In *2018 20th International Conference on Advanced Communication Technology (ICACT),* Chuncheon-si Gangwon-do, Korea (South), 2018, pp. 405–410.

[44] Lakhani A. The role of artifcial intelligence in IoT and OT security. https://www.csoonline.com/article/3317836/the-role-of-artificial-intelligen ce-in-iot-and-ot-security.html. Accessed 11 February 2020.

[45] Pal R, Adhikari D, Heyat MB, Ullah I, You Z. Yoga meets intelligent Internet of Things: Recent challenges and future directions. *Bioengineering.* 2023;10(4):459.

[46] Asif M, Khan WU, Afzal HR, Nebhen J, Ullah I, Rehman AU, Kaabar MK. Reduced-complexity LDPC decoding for next-generation IoT networks. *Wireless Communications and Mobile Computing.* 2021;2021:1–10.

[47] Waleed S, Ullah I, Khan WU, Rehman AU, Rahman T, Li S. Resource allocation of 5G network by exploiting particle swarm optimization. *Iran Journal of Computer Science.* 2021;4(3):211–219.

[48] Khan HU, Sohail M, Ali F, Nazir S, Ghadi YY, Ullah I. Prioritizing the multi-criterial features based on comparative approaches for enhancing security of IoT devices. *Physical Communication.* 2023;59:102084.

[49] Mazhar T, Irfan HM, Haq I, Ullah I, Ashraf M, Shloul TA, Ghadi YY, Imran, Elkamchouchi DH. Analysis of challenges and solutions of IoT in smart grids using AI and machine learning techniques: A review. *Electronics.* 2023;12(1):242.

[50] Pendse A. Transforming cybersecurity with AI and ML: View. https://ciso.economictimes.indiatimes.com/news/transforming-cybersecuritywith-ai-and-ml/67899197. Accessed 12 February 2020.

[51] Sethi P & Sarangi SR. Internet of things: architectures, protocols, and applications. *Journal of Electrical and Computer Engineering.* 2017; 2017:25 pages.

[52] Yaqoob I, Hashem IAT, et al. Internet of Things forensics: Recent advances, taxonomy, requirements, and open challenges. *Future Generation Computer Systems.* 2019;92:265–275.

[53] Topol EJ. High-performance medicine: The convergence of human and artificial intelligence. *Nature Medicine.* 2019;25:44–56.

[54] Al-Fuqaha A, et al. Internet of Things: A survey on enabling technologies, protocols, and applications. *IEEE Communications Surveys & Tutorials.* 2015;17:2347–2376.

Artificial Intelligence (AI)-Powered Internet of Things (IoT)

Smartening Up IoT

Amal Saif

Prince Sumaya University for Technology

Qasem Abu Al-Haija

Jordan University of Science and Technology

2.1 INTRODUCTION

Artificial Intelligence (AI) and the Internet of Things (IoT) are concepts associated in many studies, although each constitutes a pure science. IoT has become widely available, as many of the devices present in our daily lives produce vast amounts of data that are transmitted, analyzed, stored necessary, and made decisions based on these data. These devices include computers, smartphones, wearable devices, and sensors that collect environmental information, medical instruments, etc.

By 2025, the number of IoT devices will reach more than 30 billion, predicts Gartner [1]. Also, according to a McKinsey estimate, the IoT sector will make between $2.7 and $6.2 trillion in economic contributions to the world economy by 2025 [2].

As PWC mentioned, AI and IoT are the main pillars of the Fourth Industrial Revolution, improving the capability of different industry fields, increasing productivity, and decreasing costs [3]. AI with IoT is called a digital twin [4]. This twin creates smart cities with intelligent lighting and safety control systems. This twin also shows in monitoring and controlling water quality and management systems. It also exists in the healthcare sector. There are light and dark sides in the realm where AI meets IoT. This cooperation benefits

DOI: 10.1201/9781032648309-3

the economics, livelihood, production, and sustainability levels. But it also has severe security, privacy, and complexity severe issues.

This merger has increased the presence of IoT devices, as applying it to various sectors has become possible, leading to an energy drain. And for this reason, a new concept also appeared, which is known as green IoT. Green IoT aims to have an energy-efficient concern in the IoT design and to consider environmental sustainability. Many methods exist to achieve these goals, such as making the devices work with minimal energy consumption, paying attention to recycling, and depending on renewable energy [5]. This chapter provides different applications that merge these two concepts where the Internet of Things devices provide massive data being analyzed through AI, lists some challenges, and provides a brief background about these two technologies.

2.2 BACKGROUND OF AI AND IoT

IoT is the term that describes connected devices/sensors within local or world networks. It is the definition of device-device and device-human communications. The AI is the term that describes analytical methods and algorithms that can make decisions based on the trained data. There is a need for a unified definition of IoT to define the essential ideas and technologies crucial to the IoT [6]. Figure 2.1 shows the related entities and connections of the IoT, which defines IoT as connecting any devices and anyone through any network accessed at any time and place [7].

AI mimics human intelligence and behavior [8]; it consists of a subset called Machine Learning (ML). ML models need handcrafted features where human intervention is required. ML includes another subset called Deep Learning (DL) that eliminates the human intervention to extract the features; it can learn from the most important features.

AI, including ML and DL models, has produced many taxonomies. One of these taxonomies divides the AI into supervised, unsupervised, semi-supervised, and reinforcement learning. Supervised learning depends on labeled data, while unsupervised learning is the

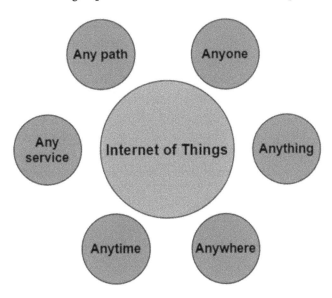

FIGURE 2.1   IoT connections and related entities. See [7].

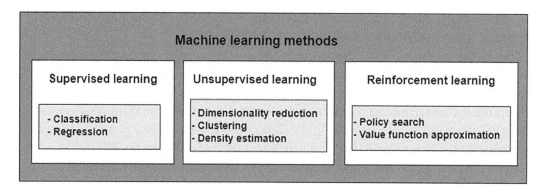

FIGURE 2.2 ML methods. See [10].

way of extracting the patterns of the data to generate clusters that include data with similar patterns. Semi-supervised learning combines the previous two methods, where partially labeled data are provided. Reinforcement learning differs from what was mentioned; this type of learning learns from trial and error [9]. Figure 2.2 shows the main categories of ML methods and their sub-methods as classified in Ref. [10].

Adaptive Convolutional Networks (ACNs), Autoencoders (AE), Generative Adversarial Networks (GANs), Convolutional Neural Networks (CNNs), Graph Neural Networks (GNNs), Recurrent Neural Networks (RNNs), and Fully connected Neural Networks (FNNs) are examples of some DL algorithms that used in IoT environments alongside other traditional ML models such as Decision Tree (DT) and Support Vector Machine (SVM) [11].

The selection of an AI model depends on the application; CNN models are used for image recognition and detection, and RNNs are used for text analysis.

This digital twin consists of two sides: the data collection from the IoT sensors and the transmission of the data from one side, and the analysis with preprocessing, integration, and making decisions from the AI side. IoT may include vision sensors; the output of these sensors is compatible with DL models that deal with object detection and image processing. One of the interesting applications is using GANs to generate new data and share the same statistics as the original data. Another type of sensor is the voice sensor, where AI models concerned with voice and text data play a core role in extracting important features and providing services based on analyzing the voice, such as Siri [12]. It is hard to mention all types of sensors that contribute to IoT; many sensors construct smart cities, smart systems, smart homes, and endless applications that turn every simple thing into smart. In general, the IoT architecture consists of three layers: the perception layer, where the sensors and actuators collect the data; the network layer, which is the layer that allows communication and transmission of the data; and the application layer, which represents where the data are being stored and analyzed [2].

2.3 APPLICATIONS OF AI IN THE IoT

2.3.1 Smart Cities

According to forecasts, 60% of the world's population will reside in metropolitan regions by 2030 due to recent migrations into these locations. As the population grows, various smart applications are released to simplify life and support smart city development [13].

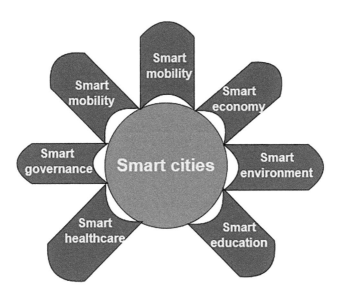

FIGURE 2.3   Smart cities. See [15].

There are many definitions of smart cities, but they all agree that smart cities raise living standards through the use of innovative technologies and provide the citizens with knowledge about what is happening around them through the communications that occur between systems to achieve high standards of economics, education, health, safety, and education. It is achieving sustainability in all areas of life, as shown in Figure 2.3. There is real competition between countries to reach the best model of smart cities [14].

2.3.1.1 Smart Libraries

AI is extensively used for prediction tasks [16]; it can preserve the borrowing time for the readers, predict the next books to reserve based on the reader's history, suggest books for readers, authorize the readers, and predict the number of copies needed to deliver to each library's branches. The IoT sensors control the borrowing-returning process without human intervention, saving time and effort compared to traditional libraries. This system helps in different cases like the pandemic of COVID-19. The IoT consists mainly of Radio Frequency Identification (RFID) technology readers and tags, WIFI access points, and BLE. At the same time, the AI fundamental keys are Natural Language Process (NLP) and other Deep Learning (DL) models [17]. Three main concepts in the smart libraries are shown in Figure 2.4.

2.3.1.2 Smart Home

AI and IoT in smart homes (AISH) exist in controlling the air-conditioning; reducing energy consumption; enhancing the home environment to make people more comfortable; and providing a caring system for older people, children, and patients at home [18]. Figure 2.5 shows a smart home model; AI can automate the device control by analyzing the collected data.

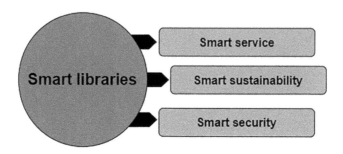

FIGURE 2.4   Smart libraries. See [17].

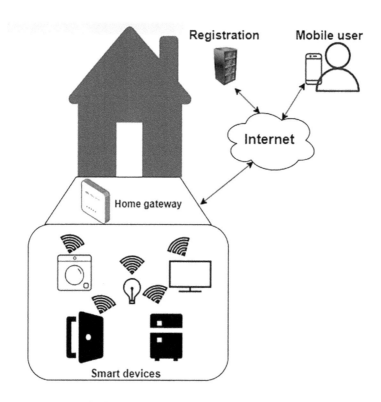

FIGURE 2.5   Smart home. See [19].

2.3.1.3 Smart Agriculture

Sensors in agricultural IoT monitor and collect plant factors, save this environment from excessive pesticide and fertilizer use that pollutes the environment, and improve production quality [20]. Frost intelligent control is done by getting data about the soil, weather, and water in the greenhouse, and frost impacts could be avoided. Climate change has demonstrated the importance of these applications and AI-IoT models and introduced fundamental changes to the greenhouse environments. Also, the frost forecast helps the greenhouse systems to provide a suitable environment for the plants [21].

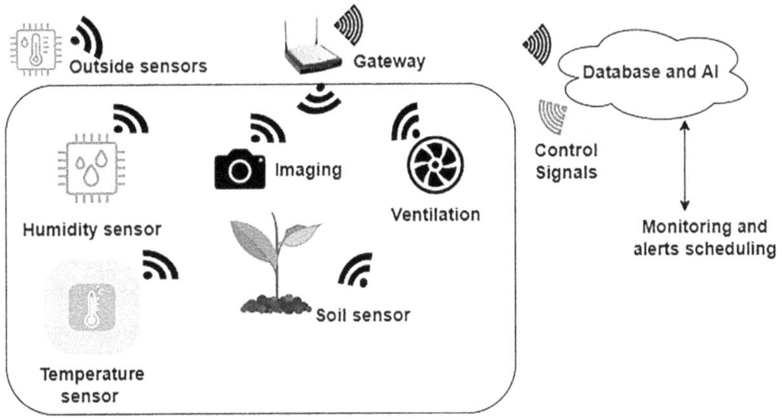

FIGURE 2.6   Smart greenhouse. See [22].

The smart greenhouse does not mean that the environment is just being monitored; according to the sensors' outputs, the AI makes decisions for the machines to control what to do to have ideal parameters of humidity, temperature, water level, and CO_2 [22], as shown in Figure 2.6.

2.3.1.4 Autonomous Vehicles

Many benefits are provided by autonomous vehicles (AVs), such as reducing accidents as humans are responsible for about 90% of crashes and reducing fuel consumption. The main challenge of this application is the data size, where sensors provide information about time, motion, other vehicles, speed, image recognition, and a long list of further details. It is a complex application that needs a high level of security. AI models handle these sensors' outputs and make decisions to control the vehicles. IoT devices are cameras, ultrasonics, radar, and sensors for long and short distances. This application faces many challenges other than complexity, such as the road and weather conditions and the liability, which make the actions harder [23].

2.3.1.5 Enterprises and Manufacturing

The small- and medium-sized enterprises (SMEs), the economic backbone, have adopted AI with IoT in their environments. For example, sensors with AI models can predict the maintenance need for machines [4]. Manufacturing intelligence is device connections to automate services to enhance productivity and reduce costs. In addition, it can provide adaptive smart decisions. Reinforcement learning is suitable for these environments where it provides dynamic plans; then the rewards are the cost, time, and equipment utilization to let the model learn the best plan [24].

2.3.1.6 Generating a Renewable Resource of Energy

IoT sensors generate energy from renewable resources. The importance of this application is reducing the demand for electricity energy, which is mainly generated from petroleum resources, and going toward clean energy resources. Using multiple sensors is more effective than a single sensor; however, to control the total generated power based on many factors like the number of sensors and the weather, there is a need for AI predictive models.

This methodology of combining these two terms in the energy generation field was proposed in Refs. [25,26]. They employed the ANN model to predict the total generated power, providing good results among root mean square error (RMSE) and R2 metrics.

2.3.1.7 Healthcare and Wearable Devices

One of the trendiest topics that merge AI with IoT is the healthcare sector. AI-IoT model is used to diagnose different diseases, where this cooperation creates the Internet of Medical Things (IoMT) concept [27]. Heart disease, diabetes, cancer, liver disorder, and many other diseases can be diagnosed accurately using AI-IoT models [28].

The data have been collected from wearable devices or instruments in the patients' rooms; the early diagnosis helps in saving human lives and giving an appropriate treatment. These devices help not only with the medical sector but also with sports. They required lightweight authentication and security algorithms [29]. Wearable devices with 5G introduce the bodyNET term [30], and the AI integration with bodyNET takes the data value to another level.

2.3.2 Security of IoT Using AI Models

As a result of valuable information provided by IoT devices, it makes them vulnerable to different attacks. AI plays an essential role in protecting and enhancing the security of the IoT. Denial of service, man in the middle, and false data injection are attacks that could occur in IoT environments. AI can be used to attack IoT devices or to enhance the security of these devices [31].

Poisoning attacks and changing the labels of the generated data targeting the AI models cause severe risks to IoT devices. CNN, ensemble learning, and graph neural networks detect suspicious behavior like malware [32]. Other approaches go beyond implementing a particular AI model for intrusion detection in IoT; they combine clustering and classification methods and conduct hundreds of experiments to enhance the performance of AI algorithms in the Intrusion Detection System (IDS) [33].

One of the interesting classifications of the AI role in IoT regarding security is what was discussed in Ref. [10]; the AI security impacts on IoT were classified as good, bad, and ugly. The good thing is using AI in combating attacks such as anomaly, misuse, and malware detection. The bad thing is exploiting the adversarial weakness in AI to attack the data, the training and testing phases, and launching white-box and black-box attacks. The ugly use is to use the AI to attack the IoT, such as evasive attacks, obtaining unauthorized access, and phishing attacks [34].

2.4 CHALLENGES AND THE FUTURE OF AI IN IoT

2.4.1 AI-IoT Challenges

Many challenges are the focus of researchers' attention. These challenges resulted from the IoT and AI separately, and new challenges resulted from the IoT-AI combination; some of them are given as follows:

1. Computation complexity is a primary challenge in developing AI models for IoT [33]. This is because of the AI models' parameters; lightweight models affect the model's accuracy. Therefore, developing a lightweight model with high accuracy is a target for many researchers [35].

2. AI should develop adaptive security models for different application scenarios.

3. Embedding AI with IoT devices to make local decisions instead of being remote. For example, some cameras have GPU cards to process the videos and images without sending them to an analytical system; this reduces the required response time and supports real-time response [36].

4. The complexity of interpreting the AI models. This is considered a big challenge for DL models; it is vital to justify the AI predictions and outputs in IoT, but things become more complex; a proposed model that embeds an explainable layer to one of the IoT datasets concluded that SHapley Additive exPlanations (SHAP) with density function analysis could enhance the AI performance by selecting the appropriate model and the important features [37].

5. Preprocessing the data collected from different resources and the common big data challenges like velocity, volume, heterogeneity, etc. [38].

6. Preserve an acceptable level of Quality of Service (QoS). Harvesting power enhances the security level; in contrast, it causes a loss in the QoS. Therefore, it is essential to have a balanced solution based on the application [39].

7. Collisions may happen due to the number of users and the lack of resources that could serve all these users.

8. Data privacy is a concern in AI-IoT models; the sensitivity of the data is different, and the diversity of the ways of dealing with the provided data by the models is also different.

9. Data liability is a complex topic, as the laws' levels and restrictions should be varied according to the application. It is only possible to list some applications, stakeholders, environments, etc., and determine who will take responsibility.

10. Provide real-time functioning; some GPUs, TPUs, and edge computing try to minimize the response time by allowing local processing and storing of the collected data.

11. Adopting AI-IoT approaches increases the need to develop proactive regulations as the pace of these technologies increases. However, the multi-stakeholders, dynamic roles, complexity of the model, diversity of the environments, many interacting factors, and the difficulties of applying similar regulations in all countries make containing all the issues more complicated [40].

As this digital twin is a promising technology for raising living standards in all fields, and many publications and articles have been discussed in detail and proposed unlimited contributions, it still has many challenges and gaps that innovative ideas can improve.

2.4.2 Future of AI-IoT

There will be a revolution in technologies in various fields, and it will directly affect people and businesses to reach a complete world of automation of things. This is clear from the rapid development of this field, as shown in Figure 2.7.

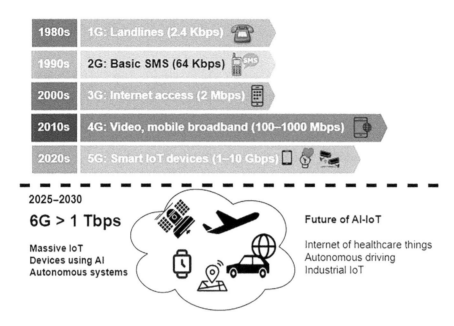

FIGURE 2.7   6G AI-IoT. See [41].

2.5 CASE STUDY

To conclude this chapter, imagine this case as an example of the use of AI and IoT, and then you will be left with open questions to think about. In a city known for a traffic jam, the government decided to take the benefits of using new technologies; they proposed an AI-IoT model to manage traffic. The system has all the required hardware and software (sensors, cameras, real-time analytical systems, traffic light control systems, etc.). Their target was to enhance the traffic flow to get a near-optimized solution.

Everything was as planned in trial, testing, and production, and they started thinking about generalizing this solution to other cities and sharing their experience. However, an unexpected event significantly disrupted the traffic flow [42].

Initial quick investigations showed that a software bug generated random decisions, the intersections suffered from chaos, and the drivers were stuck in congestion for hours. It took work to solve the problem, which needed to be clarified. This makes people lose trust in this solution, especially since their work has been delayed. After several hours, the bug was solved.

This is a simple case about the effect of mistakes in the AI-IoT model. For instance, you can project this incident onto more significant examples such as healthcare. Now think about the following questions:

1. What to do if a cyber-attack caused this disruption?

2. Who will be liable, especially since we know that negative predictions exist in any AI model?

3. How can we increase the trust in the AI-IoT models?

4. Is the human oversight of these models enough to avoid a similar situation?

5. Suggest an alternative solution if the AI model malfunctions in general.

6. Are there some sectors in which we cannot deploy AI-IoT models?

There are many aspects to consider in this era's challenges.

REFERENCES

1. H. Song, J. Bai, Y. Yi, J. Wu, and L. Liu, "Artificial intelligence enabled Internet of things: Network architecture and spectrum access," *IEEE Computational Intelligence Magazine*, vol. 15, no. 1, pp. 44–51, 2020.
2. J. Zhang and D. Tao, "Empowering things with intelligence: A survey of the progress, challenges, and opportunities in artificial intelligence of things," *IEEE Internet of Things Journal*, vol. 8, no. 10, pp. 7789–7817, 2020.
3. M. Anshari, "Workforce mapping of the fourth industrial revolution: Optimization to identity," *Journal of Physics: Conference Series*, vol. 1477, p. 072023, 2020.
4. E. B. Hansen and S. Bøgh, "Artificial intelligence and internet of things in small and medium-sized enterprises: A survey," *Journal of Manufacturing Systems*, vol. 58, pp. 362–372, 2021.
5. M. F. Tuysuz and R. Trestian, "From serendipity to sustainable green IoT: Technical, industrial and political perspective," *Computer Networks*, vol. 182, p. 107469, 2020.
6. G. S. Blair, K. Beven, R. Lamb, R. Bassett, K. Cauwenberghs, B. Hankin, G. Dean, N. Hunter, L. Edwards, V. Nundloll, et al., "Models of everywhere revisited: A technological perspective," *Environmental Modelling & Software*, vol. 122, p. 104521, 2019.
7. A. Ghasempour, "Internet of things in smart grid: Architecture, applications, services, key technologies, and challenges," *Inventions*, vol. 4, no. 1, p. 22, 2019.
8. N. A. Jebril and Q. Abu Al-Haija, "Artificial intelligence and machine learning methods in bioinformatics and medical informatics," *Emerging Technologies in Biomedical Engineering and Sustainable Telemedicine*, pp. 13–30, 2021.
9. M. M. Belal and D. M. Sundaram, "Comprehensive review on intelligent security defenses in the cloud: Taxonomy, security issues, ML/DL techniques, challenges and future trends," *Journal of King Saud University-Computer and Information Sciences*, vol. 34, no. 10, pp. 9102–9131, 2022.
10. F. Liang, W. G. Hatcher, W. Liao, W. Gao, and W. Yu, "Machine learning for security and the Internet of things: The good, the bad, and the ugly," *IEEE Access*, vol. 7, pp. 158126–158147, 2019.
11. P. Malhotra, Y. Singh, P. Anand, D. K. Bangotra, P. K. Singh, and W.-C. Hong, "Internet of things: Evolution, concerns and security challenges," *Sensors*, vol. 21, no. 5, p. 1809, 2021.
12. Z. Zhang, F. Wen, Z. Sun, X. Guo, T. He, and C. Lee, "Artificial intelligence-enabled sensing technologies in the 5g/internet of things era: From virtual reality/augmented reality to the digital twin," *Advanced Intelligent Systems*, vol. 4, no. 7, p. 2100228, 2022.
13. M. E. E. Alahi, A. Sukkuea, F. W. Tina, A. Nag, W. Kurdthongmee, K. Suwannarat, and S. C. Mukhopadhyay, "Integration of IoT-enabled technologies and artificial intelligence (AI) for smart city scenario: Recent advancements and future trends," *Sensors*, vol. 23, p. 5206, 2023.
14. C. S. Lai, Y. Jia, Z. Dong, D. Wang, Y. Tao, Q. H. Lai, R. T. Wong, A. F. Zobaa, R. Wu, and L. L. Lai, "A review of technical standards for smart cities," *Clean Technologies*, vol. 2, no. 3, pp. 290–310, 2020.
15. H. Herath and M. Mittal, "Adoption of artificial intelligence in smart cities: A comprehensive review," *International Journal of Information Management Data Insights*, vol. 2, no. 1, p. 100076, 2022.
16. Q. Abu Al-Haija, O. Mohamed, and W. Abu Elhaija, Predicting global energy demand for the next decade: A time-series model using nonlinear autoregressive neural networks. *Energy Exploration & Exploitation*, vol. 41, no. 6, pp. 1884–1898, 2023. doi:10.1177/01445987231181919.

17. S. Bi, C. Wang, J. Zhang, W. Huang, B. Wu, Y. Gong, and W. Ni, "A survey on artificial intelligence aided internet-of-things technologies in emerging smart libraries," *Sensors*, vol. 22, no. 8, p. 2991, 2022.

18. S. Sepasgozar, R. Karimi, L. Farahzadi, F. Moezzi, S. Shirowzhan, S. M. Ebrahimzadeh, F. Hui, and L. Aye, "A systematic content review of artificial intelligence and the internet of things applications in smart home," *Applied Sciences*, vol. 10, no. 9, p. 3074, 2020.

19. J. Oh, S. Yu, J. Lee, S. Son, M. Kim, and Y. Park, "A secure and lightweight authentication protocol for IoT-based smart homes," *Sensors*, vol. 21, no. 4, p. 1488, 2021.

20. X. Zhang, Z. Cao, and W. Dong, "Overview of edge computing in the agricultural internet of things: Key technologies, applications, challenges," *IEEE Access*, vol. 8, pp. 141748–141761, 2020.

21. A. Casta˜neda-Miranda and V. M. Casta˜no-Meneses, "Internet of things for smart farming and frost intelligent control in greenhouses," *Computers and Electronics in Agriculture*, vol. 176, p. 105614, 2020.

22. N. Misra, Y. Dixit, A. Al-Mallahi, M. S. Bhullar, R. Upadhyay, and A. Martynenko, "IoT, big data, and artificial intelligence in agriculture and food industry," *IEEE Internet of Things Journal*, vol. 9, no. 9, pp. 6305–6324, 2020.

23. Q. Abu Al-Haija, "SysML-based design of autonomous multi-robot cyber-physical system using smart IoT modules: A case study," In: Hemanth, D. J. (ed) *Machine Learning Techniques for Smart City Applications: Trends and Solutions. Advances in Science, Technology & Innovation*. Springer, Cham, pp. 203–219, 2022. doi:10.1007/978-3-031-08859-9_16.

24. Y. Lu, X. Xu, and L. Wang, "Smart manufacturing process and system automation – A critical review of the standards and envisioned scenarios," *Journal of Manufacturing Systems*, vol. 56, pp. 312–325, 2020.

25. V. Puri, S. Jha, R. Kumar, I. Priyadarshini, M. Abdel-Basset, M. Elhoseny, H. V. Long, et al., "A hybrid artificial intelligence and internet of things model for generation of renewable resource of energy," *IEEE Access*, vol. 7, pp. 111181–111191, 2019.

26. Q. A. Al-Haija, M. I. Al Tarayrah and H. M. Enshasy, "Time-series model for forecasting short-term future additions of renewable energy to worldwide capacity," In: *2020 International Conference on Data Analytics for Business and Industry: Way Towards a Sustainable Economy (ICDABI)*, Sakheer, Bahrain, 2020, pp. 1–6, doi:10.1109/ICDABI51230.2020.9325625.

27. R. F. Mansour, A. El Amraoui, I. Nouaouri, V. G. D´ıaz, D. Gupta, and S. Kumar, "Artificial intelligence and internet of things enabled disease diagnosis model for smart healthcare systems," *IEEE Access*, vol. 9, pp. 45137–45146, 2021.

28. A. Kishor and C. Chakraborty, "Artificial intelligence and internet of things-based healthcare 4.0 monitoring system," *Wireless Personal Communications*, vol. 127, no. 2, pp. 1615–1631, 2022.

29. D. Nahavandi, R. Alizadehsani, A. Khosravi, and U. R. Acharya, "Application of artificial intelligence in wearable devices: Opportunities and challenges," *Computer Methods and Programs in Biomedicine*, vol. 213, p. 106541, 2022.

30. Q. Shi, B. Dong, T. He, Z. Sun, J. Zhu, Z. Zhang, and C. Lee, "Progress in wearable electronics/photonics-moving toward the era of artificial intelligence and internet of things," *InfoMat*, vol. 2, no. 6, pp. 1131–1162, 2020.

31. M. Kuzlu, C. Fair, and O. Guler, "Role of artificial intelligence in the Internet of things (IoT) cybersecurity," *Discover the Internet of Things*, vol. 1, pp. 1–14, 2021.

32. H. Wu, H. Han, X. Wang, and S. Sun, "Research on artificial intelligence enhancing Internet of things security: A survey," *IEEE Access*, vol. 8, pp. 153826–153848, 2020.

33. A.A. Alsulami, Q. Abu Al-Haija, A. Tayeb, and A. Alqahtani, "An intrusion detection and classification system for IoT traffic with improved data engineering," *Applied Sciences*, vol. 12, p. 12336, 2022. doi:10.3390/app122312336.

34. V. Prutyanov, N. Melentev, D. Lopatkin, A. Menshchikov, and A. Somov, "Developing IoT devices empowered by artificial intelligence: Experimental study," In: *2019 Global IoT Summit (GIoTS)*, Aarhus, Denmark, 2019, pp. 1–6, IEEE.

35. H. Wang, L. Li, Y. Cui, N. Wang, F. Shen, and T. Wei, "Mbsnn: A multi-branch scalable neural network for resource-constrained IoT devices," *Journal of Systems Architecture*, vol. 142, p. 102931, 2023.
36. Z. Lv, L. Qiao, S. Verma, and Kavita, "AI-enabled iot-edge data analytics for connected living," *ACM Transactions on Internet Technology*, vol. 21, no. 4, pp. 1–20, 2021.
37. R. Younisse, A. Ahmad, and Q. Abu Al-Haija, "Explaining intrusion detection-based convolutional neural networks using Shapley additive explanations (SHAP)," *Big Data and Cognitive Computing*, vol. 6, no. 4, p. 126, 2022.
38. M. Younan, E. H. Houssein, M. Elhoseny, and A. A. Ali, "Challenges and recommended technologies for the industrial Internet of things: A comprehensive review," *Measurement*, vol. 151, p. 107198, 2020.
39. B. Mao, Y. Kawamoto, and N. Kato, "AI-based joint optimization of QoS and security for 6g energy harvesting Internet of things," *IEEE Internet of Things Journal*, vol. 7, no. 8, pp. 7032–7042, 2020.
40. S. Hadzovic, S. Mrdovic, and M. Radonjic, "A path towards an internet of things and artificial intelligence regulatory framework," *IEEE Communications Magazine*, vol. 61, no. 7, pp. 90–96, 2023.
41. D. C. Nguyen, M. Ding, P. N. Pathirana, A. Seneviratne, J. Li, D. Niyato, O. Dobre, and H. V. Poor, "6g Internet of things: A comprehensive survey," *IEEE Internet of Things Journal*, vol. 9, no. 1, pp. 359–383, 2021.
42. Q. A. Al-Haija and N. A. Jebril, "ARM cortex based modeling and implementation of a self-controlled traffic light system," In: *IET Conference Proceedings*, pp. 607–612, 2021, doi:10.1049/icp.2021.0942.

CHAPTER 3

Bridging the Future

The Confluence of Internet of Things and Artificial Intelligence in Communication System

Rahim Khan and Xuefei Ma

Harbin Engineering University

Sher Taj

Northeastern University

Daqing Normal University

Hina Hassan

Harbin Normal University

Inam Ullah

Gachon University

Abdullah Alwabli

Umm Al-Qura University

Yuning Tao

South China University of Technology

Habib Ullah

Nanjing University of Aeronautics and Astronautics

DOI: 10.1201/9781032648309-4

3.1 INTRODUCTION

In the ever-evolving landscape of technology, two groundbreaking innovations have emerged as catalysts for transformative change: AI and IoT [1]. These two technological giants have not only redefined the way we connect and communicate but also paved the way for a future where our devices, systems, and environments become more intelligent and interconnected than ever.

In this chapter, we embark on a journey into the fascinating world of IoT and AI, exploring their individual realms and uncovering the profound impact they have on communication systems. We will delve into the fundamental concepts, mechanics, and applications of both IoT and AI, shedding light on how they work in tandem to revolutionize the way we monitor, control, and interact with the world around us.

From the foundational principles of IoT and its ecosystem components to the intricate workings of AI, we will unravel the complexities of these technologies. We will discuss the myriad benefits and challenges that arise from their convergence, as well as the boundless possibilities they offer for shaping our future. As we navigate through this chapter, you will gain a comprehensive understanding of not only what IoT and AI are but also how they synergize to create AI-aided IoT technologies.

3.2 THE EVOLUTION OF INTERNET OF THINGS

The Internet has profoundly influenced how we live, communicate, connect, travel, and look after ourselves. Many modern household devices and gadgets, capable of online communication and interaction, include vacuum cleaners, washing machines, and more. These devices can be monitored and managed remotely using smartphone apps. This burgeoning field is termed the IoT. IoT settings, as mentioned in Ref. [1], comprise smart home products and services that communicate with one another in real time. Governments worldwide, including those of Japan, the UK, and the USA, have poured substantial funds into IoT research and development, underscoring their commitment to this transformative technology [2]. IoT has the potential not only to transform the lives of everyday people but also to redefine the global economic landscape, affecting governments, industries, and service sectors.

A primary aspiration of consumer IoT is enhancing daily convenience. For instance, nobody likes having to leave their comfort zone to turn off the TV or a lamp. Smart home solutions that are IoT-powered automate these processes. These intelligent homes can adapt to user preferences, whether it's adjusting lights, regulating room temperature, or playing music, through sensor-based technology. As devices evolve, the smart home can integrate these advanced features, like setting laundry times in intelligent washing machines. When individuals are away, soil humidity and ambient temperature sensors can ensure plants are watered adequately. If someone is about to leave home, smart cars can sense this and adjust their settings accordingly.

However, IoT's potential stretches beyond individual convenience. The next IoT development is anticipated to give rise to smart cities and countries. To improve energy conservation, traffic control, and planning for cities, data from homes, communities, and institutions may be combined at the city level. This approach will assist the general public and offer additional advantages. In the corporate realm, this translates to streamlined operations, automation, efficient resource utilization, and enhanced customer experiences.

Yet, with these advancements come significant concerns, primarily about security and privacy. Our devices often transmit information to external entities, including home security providers, energy suppliers, and vehicle makers. Online, smart assistants analyze our spoken instructions, which could potentially lead to data being compromised. Such data, if misappropriated, could compromise our security and privacy. Questions about the data type and volume collected, and who has access to it, remain pertinent. Each improperly managed device can become a security vulnerability, regardless of its user—be it an individual, a business, or a government entity. All stakeholders need to proactively address these security concerns.

Many are unaware of the depth and breadth of data collection by their smart devices. This finding serves as the main argument of my thesis, which seeks to answer questions such as: How does your smart TV observe your activities? How might hackers exploit security weaknesses in everyday items like light bulbs? Or why might your heating system request your phone number, potentially relaying it to marketers? It's high time we prioritize security, privacy, and safety. Consumers should be discerning about the smart devices they purchase and the paramount considerations when using them.

In conclusion, manufacturers bear a significant responsibility. As connected devices become prime targets for hackers and cybercriminals, manufacturers must ensure the security of their products, routinely update them, and empower consumers to counteract the ongoing threats posed by malicious cyber activities.

3.3 WHAT IS THE IoT?

It is important to first define the IoT before delving into its core elements. Broadly stated, the IoT can be defined as an intricate web of physical entities empowered with the following:

- **Sensors:** These gather data.
- **Identifiers:** They pinpoint the data's origin, such as specific sensors or devices.
- **Software:** This is used for data analysis.
- **Internet Connectivity:** This ensures communication and alerts.

IoT is essentially a network of uniquely identifiable devices that are connected to the internet everywhere. This network allows these objects to relay information to manufacturers, operators, and other interconnected devices using the Internet's telecommunications infrastructure. It enables tangible objects to relay specific data and to be remotely managed over the Internet. This facilitates a more seamless merger between our physical environment and digital systems, enhancing efficiency, precision, and economic advantage. Every object is distinctively identified through its in-built computing mechanism, enabling it to function within the broader Internet framework.

There is a consensus among businesses and tech experts about the exponential growth in the number of IoT-connected items. Gartner envisaged a future where the operational count of devices would reach a staggering 20 billion by 2020. On the other hand, Cisco's

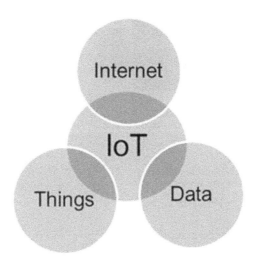

FIGURE 3.1   Basic definition of IoT.

estimation was even more ambitious, predicting a whopping 26.3 billion devices to be in existence within the same time frame. These projections encompassed a wide range of gadgets, including tablets, televisions, smartphones, computers, and various other items that are connected to the online realm. Some argue these numbers are understated, speculating that soon, any object equipped with a basic microcontroller, rudimentary switch, or even a QR code could be Internet-connected. Such a perspective is further supported by the renowned Moore's Law, which astutely observes that the number of transistors on an intricate integrated circuit tends to double approximately every 18 months. Figure 3.1 shows the basic definition of IoT.

At its essence, the IoT strives to interconnect nearly all entities—encompassing sensors, gadgets, contraptions, individuals, creatures, and even foliage—via the World Wide Web, with the primary intent of monitoring and/or overseeing. These connections are not just informational; they are tangible, allowing users to access and control these "things" as needed. Therefore, the mere act of connecting objects is not the end goal. The true objective is extracting insights from these connected entities to enhance products and services.

3.4 HOW TO REMOTELY MONITOR AND CONTROL ITEMS WORLDWIDE?

Let us embark on this journey by delving into the primary inquiry. The essential prerequisites for the IoT encompass an individualized identifier for every entity in question (such as an IP address), the capability for these entities to engage in discourse with one another (as exemplified by wireless communications), and the aptitude to ascertain specific particulars about said entities by means of sensors. With these three constituents firmly established, one can scrutinize these "entities" from any corner of the world. Additionally, a vital foundational need is a means of communication, typically managed by a telecommunications network. Figure 3.2 illustrates the essential components of an IoT solution.

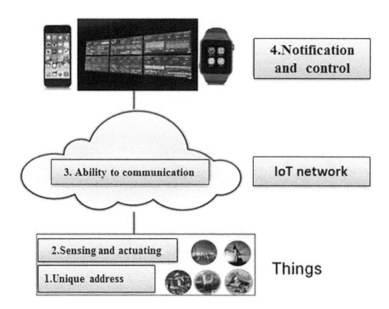

FIGURE 3.2 Fundamental prerequisites of an IoT resolution.

3.5 WHY DO WE SEEK TO OVERSEE AND REGULATE THINGS?

There exist diverse incentives for remotely supervising and regulating objects through the Internet. Some justifications encompass granting specialists the ability to oversee factors like an individual's body temperature or blood pressure from the comfort of their own homes; attaining wisdom by directing a smartphone towards a fascinating object; requesting information that current search engines, such as Google, do not provide (for instance, the whereabouts of one's vehicle keys); empowering authorities to efficiently manage resources in intelligent cities, such as energy or the granting of driver's licenses and other Department of Motor Vehicles documents for senior citizens; and presenting cost-efficient entertainment and games for both juveniles and elders. These circumstances epitomize substantial commercial and service prospects that can amplify the economic influence for consumers, enterprises, governments, medical facilities, and various other establishments.

3.6 WHO WILL BE IN CHARGE OF MONITORING AND MANAGEMENT?

Usually, both individuals and machines can monitor and control IoT services. For instance, a homeowner might use a mobile device to monitor their house via a security system they have installed and set up. In the realm of possibilities, individuals possess the capability of exerting their influence over various aspects, such as illuminating radiance, initiating the atmospheric cooling process, or ceasing the operation of the warmth-inducing mechanism. Conversely, in a different scenario, a designated entity could be entrusted with the supervision and administration of essential amenities for patrons from a central hub of network operations, as visually depicted in the embodiment showcased in Figure 3.3.

It is crucial to emphasize the importance of security in this context. Unauthorized access must be prevented, especially from malicious hackers who could potentially deceive

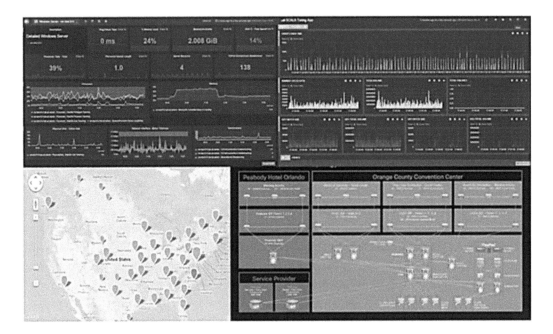

FIGURE 3.3   Representation of network operations center monitoring systems.

homeowners by showing them outdated footage while a burglary is underway. For sectors with high stakes, such as patient health monitoring or banking, the need for tight control is even more pronounced.

3.7 UNDERSTANDING THE FUNCTIONALITY OF THE IoT

The digital realm has revolutionized the worldwide scenery and the manner in which communication and labor are carried out. This progression is destined to persist with the advent of cutting-edge technologies such as 5G and groundbreaking digital protocols like Li-Fi. The IoT enhances this interconnectedness by facilitating numerous devices to synchronize concurrently via the internet, fostering not only human-to-machine but also M2M communications [3–5]. Such capabilities have unlocked a myriad of opportunities at both personal and professional levels.

Understanding and leveraging the IoT might seem daunting, but its complexity largely depends on the technological prowess of the intended audience. It is anticipated that younger individuals and families will be more inclined to harness its benefits compared to the elderly [6–8].

In a basic IoT setup, devices equipped with sensors connect to IoT platforms. These platforms gather, share, and assimilate data from various devices, subsequently employing intelligent analytics to convey crucial details to specific applications designed to meet particular requirements [9,10]. These advanced IoT platforms discern the essential data from the irrelevant. The acquired data serve myriad functions, such as pattern recognition, offering suggestions, pinpointing potential issues, and facilitating informed decision-making [11,12]. Figure 3.4 displays the functionality of IoT.

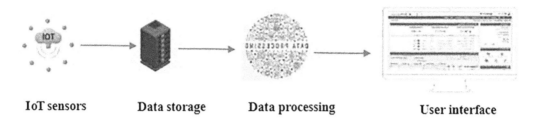

IoT sensors	**Data storage**	**Data processing**	**User interface**

FIGURE 3.4   The functionality of IoT.

For example, consider owning a sports clothing and accessory business and wanting to ascertain the popularity of various optional sports gear, such as fishing equipment, skating gear, or skiing accessories. Through IoT solutions, sensors can be employed to monitor which sections of the store attract the most traffic and where customers linger the longest. Leveraging these data, one can refine the business approach, identify top-selling items, ensure popular products remain in stock, and optimize both time and resources.

The reach of IoT is vast, spanning numerous economic sectors. Its commercial uses span the gamut, from household gadgets to transportation to factories to hospitals to banks to the military [13]. Moreover, IoT systems can integrate AI and machine learning (ML) to enhance and streamline data collection.

3.8 ELEMENTS WITHIN THE IoT ECO SYSTEM

Integration of several technologies—AI, data analytics, sensors, network access, and a graphical user interface—forms the foundation of the IoT ecosystem. Let us delve into them individually:

3.8.1 Devices

IoT devices encompass various hardware forms such as sensors, gadgets, household appliances, and machines designed for specific tasks and with the capability to transmit data online. Mobile phones, factory gear, military hardware, healthcare gadgets, environmental sensors, and more might all benefit from using these components [14]. Consequently, these devices find their application in a wide range of products, from lights, refrigerators, security systems, and printers to mobile phones, washing machines, headphones, and wearable gadgets. Their complexity varies, and they can gather and disseminate data, facilitated by the affordability of computer chips and the existence of swift wireless networks.

3.8.2 Sensors

Sensors are an indispensable part of the IoT infrastructure. They have the capability to detect and observe changes in their surroundings and convert these observations into signals comprehensible to both humans and machines [15]. There are two main types of sensors: active and passive, as well as analog and digital. Common types of sensors used in the IoT include thermometers, accelerometers, gyroscopes, gas analyzers, hygrometers, barometers, light detectors, and infrared cameras. Their pivotal role enhances operational efficacy, reduces costs, and boosts worker safety and productivity [16].

3.8.3 Connectivity

The essence of IoT devices lies in their ability to connect via the internet. The scale of these networks can be adapted based on the size and reach of the IoT system [17]:

- **LAN (Local Area Network):** This refers to a collection of devices interconnected at a single physical point, such as a home, office, or building. A LAN can range from a single-user home configuration to thousands of users, as well as gadgets in a large-scale business network in an institutional context.

- **PAN:** This network facilitates data transfer among devices in close proximity to an individual. It usually involves wireless communication between smartphones, PCs, tablets, etc., aiming for data exchange between the devices or to a central server. The IEEE 802.15 group predominantly oversees advancements in PANs.

- **MAN (Metropolitan Area Network):** This network bridges computers across a metropolitan region, which might include a major city or a cluster of smaller cities and towns. A MAN is characterized by a larger coverage area compared to a LAN but is still more limited in scope when compared to a WAN.

- **Wide Area Network (WAN):** It comprises interconnected LANs or other networks. In essence, a WAN is a network that is comprised of several interconnected networks, with the Internet serving as the most prominent and expansive manifestation of a WAN on a worldwide scale.

3.8.4 Artificial Intelligence

Individually, IoT and AI stand as immensely potent technologies. However, when integrated, their efficacy is magnified. AI is typically defined as a system's ability to execute tasks or intelligently analyze and learn from data. When AI converges with IoT technology, it results in a device capable of analyzing data and making informed decisions autonomously, encapsulating the core ideas of IoT. Figure 3.5 shows the fundamental building blocks of an IoT setup.

3.9 RELATED TECHNOLOGIES FOR THE IoT

Internet Protocol version 6; WSN; RFID devices; cloud computing; near-field communication systems; service-oriented architectures; global positioning systems; third-, fourth-,

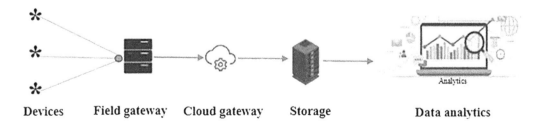

Devices   Field gateway   Cloud gateway   Storage   Data analytics

FIGURE 3.5   Fundamental structure of an IoT system.

FIGURE 3.6   Principal technologies and protocols enabling IoT systems.

and fifth-generation mobile networks; and geographic information systems are all essential to the smooth functioning of the IoT. Of these technologies, IPv6, RFID, and WSN are deemed fundamental for the optimal functioning of the IoT. Figure 3.6 showcases various technologies implemented in different IoT systems.

3.10 CHALLENGES OF INTERNET OF THINGS

IoT is undoubtedly a captivating topic of discussion. Sophisticated sensors, ultra-low-power microcontrollers, and wireless technologies have all contributed to making IoT a reality, but its broad acceptance is still an open question. Despite these advancements, several key issues prevent the full realization of an IoT-dominated world.

3.10.1 Connectivity

The global reach of the internet has expanded, yet many remote areas and small villages remain offline. Companies like Google have explored innovative solutions like balloon-powered internet, but universal internet coverage remains elusive. A foundational assumption of IoT is the availability of consistent, fast, and reliable network connectivity, which is currently one of the biggest hurdles.

3.10.2 Security and Trust

The twin challenges of trust and security are significant barriers to the widespread acceptance of IoT. Users harbor legitimate concerns about the safety of sharing their data in an interconnected world. The interlinking of information and devices creates potential vulnerabilities. For instance, an IoT-connected home might be more susceptible to burglaries, or a company might face data breaches with competitors accessing its production data. Despite robust security measures, issues surrounding trust persist.

3.10.3 Interoperability

Establishing meaningful connections between a myriad of devices is challenging. For seamless functioning, IoT demands standardized platforms that ensure connectivity, remote operability, cross-device programmability, and independence from specific models, manufacturers, or industries. Essentially, IoT should be platform-agnostic and compatible with varied operating systems, OEMs, connectors, versions, and protocol standards.

3.10.4 Energy and Environment

Many contemporary devices are battery-operated with limited lifespans. Anticipated popularity growth in IoT will lead to a significant increase in both the quantity of devices and the scale of the network. Relying on current energy sources will be untenable for powering this expansive network. A shift toward alternative energy sources will become crucial.

Additionally, if upcoming devices maintain short lifespans akin to current devices, the resulting electronic waste will be monumental. This would not only disrupt environmental equilibrium but also pose significant hazards. Hence, the evolution of IoT should emphasize environmentally friendly designs. This entails research into alternative energy sources, creating longer-lasting devices and focusing on the reusability and recycling of materials.

3.11 FUTURE APPLICATION AREAS OF INTERNET OF THINGS

The IoT is still in its developmental phase, hindered by various factors that prevent its full utilization. However, the horizon for IoT is expansive, encompassing technological advancements and their anticipated applications. Key initiatives in this domain could significantly propel the growth of this rapidly evolving technology. Here are some potential applications for the IoT:

3.11.1 Agriculture

With the escalating need for food, there is immense strain on agricultural production. IoT can enhance agricultural yield and productivity, as well as modernize storage and distribution methods. In developing nations, high initial investment costs often hinder growth in this sector. M2M or IoT solutions can alleviate these challenges. Presently, developed regions like North America are leading in satellite-based M2M/IoT adoption, while regions like the Middle East and Africa anticipate a 5.2% annual growth. In Asia, agriculture's potential growth could double revenues by 2023, raising its global share from 17.6% in 2013 to 23.2% in 2023.

3.11.2 Construction Industry

IoT in construction provides comprehensive structural details, including imagery, over time. It aids in identifying hazards posed by environmental factors, earthquakes, engineering efforts, landslides, and industrial activities. IoT surveillance reduces the need for frequent site visits. Its adoption in this sector is expected to witness a 25% annual growth, reaching $3.4 billion by 2022.

3.11.3 Healthcare

In developing nations, dropping sensor costs will soon amplify IoT's advantages. These reduced expenses will foster the increased application of IoT in healthcare, exemplified by innovations like smart bandages. These bandages, with integrated sensors, can notify both patients and doctors about healing complications. Very Small Aperture Terminal networks are also essential, acting as a fallback for terrestrial networks, especially vital in healthcare in developing regions. The adoption of IoT tools, such as wearables and remote monitoring instruments, will surge, particularly after the COVID-19 pandemic, which hastened tele-health efforts.

3.11.4 Geospatial Data Mining

Certain IoT applications demand geospatial analysis data. Geospatial mapping is increasingly used for urban mapping, including traffic, pollution, and specific areas, as well as disaster monitoring like earthquakes and landslides. Potential applications include developing urban risk maps through crowdsensing from mobile devices or real-time urban sensing for personalized travel guidance using data mining and geospatial techniques.

In the realm of AI, gleaning valuable insights from intricate environments at varying spatial and temporal resolutions remains a tough research topic. Contemporary techniques employ shallow learning combined with both supervised and unsupervised methods. The field of ML needs to shift its focus toward deep learning (DL), which seeks to understand multiple abstraction layers to interpret data accurately. Additionally, resource limitations in sensor networks present chances to integrate DL.

3.11.5 GIS-Based Visualization

Advancements in display technologies have paved the way for innovative visualization techniques. The progression in this technology has led to enhanced data portrayal via touch-screen interfaces, enabling users to navigate information more effectively and swiftly. The introduction of 3D displays promises greater potential for research in visualizing processes or events. Yet, data sourced from ubiquitous computing are not always primed for immediate visualization; they often require additional processing, especially for diverse spatio-temporal data. There is a need to devise new visualization strategies to present data from varied sensors within a 3D environment that also changes over time. Another intricacy in visualizing IoT-collected data is its geo-relevance and sparse distribution, suggesting a need for an Internet GIS-based framework to address these challenges.

3.11.6 Sensor Fusion

The growth in sensor fusion has unveiled new possibilities, notably in domains like smart homes and healthcare. Nonetheless, such applications can pose substantial privacy issues, which are critical for understanding IoT management. Platforms developed for sensor fusion have the potential to introduce unprecedented services. For instance, sensors placed on cardboard boxes for fruits and vegetables can monitor location, temperature, and movements. They can even detect the freshness of the produce and provide early alerts on potential spoilages.

IoT evolution demands that entities such as environments, cities, buildings, vehicles, wearables, and mobile gadgets continually accrue associated data. This enables them to generate novel information. The emergence of 5G is anticipated to amplify data transmission speeds and bolster real-time analytics, thereby enhancing the efficiency and efficacy of IoT devices.

3.12 ARTIFICIAL INTELLIGENCE OF THINGS (AIoT)

The term "smart" captivates us, yet current advancements do not quite embody human-like intelligence. Take smartphones as an example. Even though they are labeled "smart," they do not perform many tasks autonomously. Ideally, a smartphone should mute notifications when the user is driving to minimize distractions. Achieving this would necessitate a connection between the individual, the smartphone, and the vehicle. Similarly, if the user falls ill, the smartphone should be capable of placing an emergency call to a relative or nearby hospital. To facilitate such functionalities, the device would need access to relevant data and connections. If we delve into more examples, it is evident that for objects to truly function smartly, everything in the physical world needs to be interconnected. Achieving genuine "smartness" hinges on the capabilities of AI.

AI is poised to endow machines with human-like thinking, propelling industries into a new digital era. Whether it is humans, fauna, flora, machinery, household gadgets, or natural elements—networking them and enabling intelligent decision-making can lead to a more automated environment. Real autonomy of the physical world necessitates ML [18], which imitates human learning processes, and a data analysis (DA) [19] component. While ML develops methods for self-reliance and automation, DA interprets historical data to refine future actions. The momentum behind integrating ML and DA into smart systems, including sensors [20] and embedded systems [21], is palpable. AI's underlying technology is mesmerizing, prompting us to reevaluate our understanding of life's essence and our professional roles. The rapid evolution driven by ML and DA in the realm of AI highlights the necessity to examine its trajectory, challenges, and potential risks.

Central to this wave of change is the IoT [22–24], envisioning a world teeming with interconnected smart devices or smart objects [25–27]. These connections span human-to-human, human-to-object, and object-to-object interactions. The Internet of Everything [28] expands on this, proposing a network where every entity, whether animate, inanimate, or digital, is linked. When implemented, the result is a cyber-physical system [29]. This interconnected ecosystem is a data goldmine, paving the way for extracting insights. Multiple fields, including database management systems [30], pattern recognition [31], data mining [19], ML [18], and big data analytics [32–34], will need to refine their techniques to manage this influx of information.

3.13 ARTIFICIAL INTELLIGENCE

AI involves the creation of machines designed to mimic human intelligence, enabling them to perform tasks that have traditionally been the purview of the human mind. Applications, processing speed, flexibility, and usefulness are just a few of the areas where the reach and potential of AI-driven systems are growing quickly. As they improve in capability, robots are taking over an increasing number of previously human-only activities. While humans inherently "make" perfect decisions based on context, AI systems merely "select" the best decision for a given moment. In essence, the nuanced creativity in human decision-making is absent in AI. Although human creativity will inevitably redefine productive roles, AI systems have proficiently minimized redundant human tasks and delivered faster results.

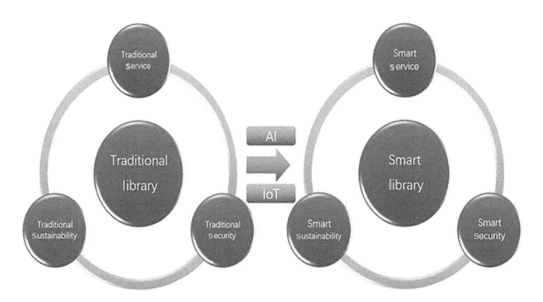

FIGURE 3.7 Evolution from conventional libraries to intelligent libraries through the implementation of AI and IoT.

Presently, most AI initiatives fall under "Narrow AI," which enhances specific tasks. Yet, the aspiration is to surpass this limitation.

The development of AI is being driven by a convergence of fields such as computing, philosophy, mathematics, biology, statistics, psychology, physics, and sociology. These fields collaborate to accentuate AI's interdisciplinary character. The essence of intelligence lies in the vast data emerging from these domains. Analyzing these data to discern under-lying principles is crucial. While the human brain can achieve this, it is often a lengthy process due to the challenging characteristics of real-world data, such as its immense vol-ume, unstructured format, diverse origins, the requirement for real-time processing, and constant flux. Other nuances include volatility and virality.

In essence, AI represents a sophisticated approach to effectively harness these data, making it comprehensible, amendable (especially when inaccuracies arise), relevant, and insightful. Central to AI's prowess is its dependency on data science techniques. Broadly speaking, data science crafts tools and methodologies to scrutinize vast datasets and extract valuable insights. Thus, data science emerges as a convergence of multiple research areas.

In contrast to conventional libraries, individuals can primarily experience the advan-tages of library advancement propelled by AI and IoT across three main areas: intelligent services, sustainable smart solutions, and enhanced security, as illustrated in Figure 3.7.

3.14 HOW DOES ARTIFICIAL INTELLIGENCE OF THINGS WORK?

In the realm of the Artificial Intelligence of Things (AIoT), AI is integrated into infrastruc-ture elements such as programs and chipsets, all of which are linked via IoT networks. Application programming interfaces ensure seamless operation and communication between all hardware, software, and platform components, eliminating any hassles for the end user. When active, IoT devices generate and collect data, which AI then evaluates to offer insights, enhancing efficiency and productivity. These insights are derived through techniques like data learning.

AIoT systems can be classified into two primary categories: cloud-based and edge-based.

3.14.1 Cloud-Based AIoT

Often referred to as the IoT cloud, this approach manages and processes data from IoT devices through CC platforms. It is crucial to link IoT devices to the cloud as it serves as the central hub for data storage, processing, and accessibility for various applications and services. Figure 3.8 illustrates the implementation of AIoT architecture in a cloud-based environment. The cloud-based AIoT is structured into four main layers:

1. **Device Layer:** This layer encompasses diverse hardware types such as tags, beacons, sensors, vehicles, industrial equipment, embedded systems, and health and fitness gadgets.

2. **Connectivity Layer:** This segment includes field and cloud gateways, which can be either hardware or software components, connecting cloud storage to controllers, sensors, and other smart devices.

3. **Cloud Layer:** This layer is responsible for data processing through an AI engine, data storage, visualization, analytics, and data accessibility via an API.

4. **User Communication Layer:** This section involves web portals and mobile apps.

3.14.2 Edge-Based AIoT

Here, data from IoT devices are processed close to the source, which reduces the bandwidth needed for data transfer and prevents potential analysis delays. Figure 3.9 shows

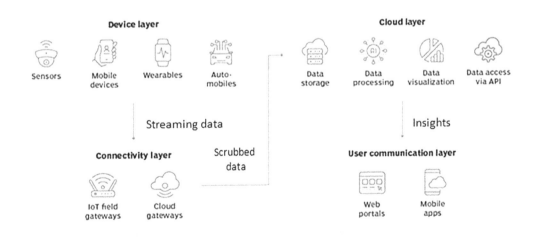

FIGURE 3.8   Implementation of AIoT architecture in a cloud-based environment.

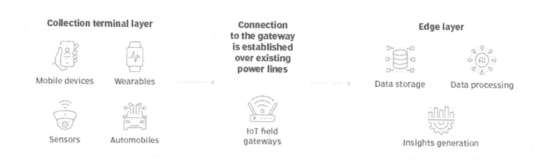

FIGURE 3.9   Implementing AIoT architecture at the edge.

the implementation of AIoT architecture at the edge. The edge-based AIoT is divided into three key layers:

1. **Collection Terminal Layer:** This layer spans a variety of hardware devices such as embedded systems, vehicles, industrial machinery, tags, beacons, sensors, mobility gadgets, and health and fitness devices, all linked to the gateway over existing power lines.

2. **Connectivity Layer:** This layer includes the field gateways to which the collection terminal layer connects over existing power lines.

3. **Edge Layer:** This part provides facilities for data storage, processing, and generating insights.

3.15 WHAT ARE THE BENEFITS AND CHALLENGES OF AIoT?

The benefits of AIoT encompass the following:

- **Operational Efficiency:** IoT devices with AI capabilities can interpret data to identify patterns and insights, allowing systems to adjust operations for greater efficiency.

- **Real-Time Adjustments:** Systems can produce and evaluate data to pinpoint failures and make necessary modifications immediately.

- **Data Analysis Efficiency:** Less human intervention is needed to monitor IoT devices, leading to cost savings.

- **Scalability:** The number of devices within an IoT framework can be augmented to refine existing processes or add new functionalities.

- **Innovative Technology Synergy:** AIoT is revolutionary, providing mutual advantages for both AI and IoT. IoT improves AI by increasing connection and data exchange, whereas AI enhances ML and decision-making in IoT. Enhanced value extraction from IoT data is one way in which this consolidation might help businesses improve their products and services.

- **Improved Security:** While IoT devices might be vulnerable to security threats, AI can pinpoint and counteract these threats by analyzing sensor data to detect anomalies and possible security infringements.

- **Minimized Human Errors:** Businesses suffer significant losses annually due to human mistakes. Integrating ML with IoT can substantially reduce these errors. As data flow through various stages or locations, it's prone to errors, such as incorrect data entry. AIoT examines data at its origin, decreasing data transit and intermediaries, and subsequently, the likelihood of mistakes.

- **Personalization:** AI can utilize data collected by IoT devices about user preferences and behaviors to enhance user experiences. For instance, an intelligent speaker can ascertain a user's music tastes and autonomously curate personalized playlists.

However, AIoT is not without potential pitfalls. For instance, a malfunctioning autonomous delivery robot could delay product deliveries. Smart retail outlets might misread a customer's facial features, leading to unintentional shoplifting, or a self-driving car might overlook an approaching stop sign, resulting in an accident.

AIoT also presents several challenges:

- **Cybersecurity Concerns:** The expansion of AIoT-connected devices amplifies the threat of cyberattacks and security breaches.

- **Integration Complexity:** Merging IoT with AI can be intricate, requiring specialized knowledge and expertise.

- **Data Management Issues:** Robust data management strategies are needed to process the diverse data collected by sensors.

- **Expense:** Implementing AIoT technologies often requires specialized hardware, software, and skilled personnel, leading to high costs.

- **Privacy Risks:** Questions arise about the handling and storage of data collected by AIoT devices, potentially leading to privacy infringements and legal ramifications.

3.16 IoT TECHNOLOGIES ENHANCED BY AI

Both IoT and AI have significantly transformed the field of librarianship. Yet, depending solely on a single technology doesn't maximize its potential in real-world applications.

From a service standpoint, the introduction of RFID-based self-checkout machines has undoubtedly increased efficiency compared to traditional librarian tasks. However, if these machines are placed in less-than-optimal locations, their efficiency can be reduced due to underutilization. Leveraging AI can assist in determining the best locations for these machines by analyzing past usage data and patterns of readers using AI algorithms.

In terms of sustainability, having high levels of light in a reading room corner where books are seldom accessed is wasteful. By analyzing previous data on reader motion patterns, AI can intelligently change the brightness. In addition, many air conditioners, which use a lot of power, are installed in study nooks, data centers, and compact stacks. Overusing these devices in low-traffic areas is inefficient and wastes resources. This issue can be addressed by either placing the right number of air conditioners or dynamically adjusting their operational modes, informed by historical environmental data.

Regarding security, potential malicious borrowing activities can be preemptively identified through AI's analysis of past borrowing data collected from RFID self-checkout machines. This can then initiate further verification steps, such as confirming alert messages from the AI system. Generally, AI can swiftly detect anomalies when there's a noticeable departure from standard operations.

However, solely relying on AI has its limitations. For instance, the RFID monitoring system might alert staff if someone with bad intentions tries to leave the library with stolen books in their luggage. Moreover, without IoT devices, it's impossible to gather and analyze the reader data essential for AI algorithms [35–38].

The benefits of both AI and IoT in the setting of a Smart Library can be amplified by the other. A holistic exploration of how AI supports IoT in the "Smart Library" is approached from three distinct perspectives.

3.16.1 Library Service Applications with Intelligent Features

Librarians work to provide a welcoming space for their patrons and utilize the library's limited resources. Traditional methods have not fully addressed the issue of seat arrangement. For instance, a vacant yet reserved seat, especially during exam times, represents a significant resource wastage. Newer solutions using AI-powered IoT are being introduced to tackle these challenges. The work in Refs. [39,40] shows a system where seat usage can be dynamically tracked and managed using web applications, pressure, and RFID sensors. In Refs. [41,42], users can easily manage reading room seats using a mobile device, adjusting based on real-time data. This concept is also applied in study room management, as illustrated in Ref. [43]. Here, a step-by-step process employs facial recognition and sensors to manage reservations. A block diagram representing this study room occupancy procedure is shown in Figure 3.10. Intriguingly, a network of BLE beacons, KNN algorithm, and Wi-Fi may be used to track the changing occupancy of a given area [44]. This enables

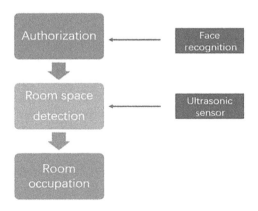

FIGURE 3.10   A diagram depicting how the classroom's available seats are assigned and used.

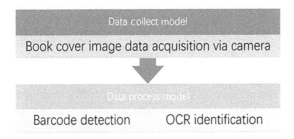

FIGURE 3.11   Depicting the flow chart of the personalized activities learning system.

students to form study groups based on real-time location and interests, thus optimizing space and enhancing smart service efficiency.

Book sorting is crucial for enhancing the efficacy of the book circulation service. Barcode and DL-powered optical character recognition (OCR) work together to dramatically improve productivity [45]. Figure 3.11 depicts the organizational structure of the system, including the components of the Data Collection Model as well as the Data Process Model. In Ref. [46], a drone robot equipped with visual localization and OCR assists with book inventory tasks. The setup in Refs. [47–49] uses CC and recommendation systems to suggest books to students based on their reading history, saving time and costs. For the visually impaired, the solution in Refs. [50–52] uses OCR, deep learning, and sensors to convert physical book content into audio. Lastly, Ref. [53] introduces an innovative model for online teaching. Using real-time class interaction data and ML, the system updates cloud-stored data, offering tailored evaluations of students' cognitive features.

In Ref. [54], a set of solutions is presented to address the issue of incorrectly reading RFID tags during the circulation process. By using prior received signal strength (RSS) measurements combined with KNN, it is possible to accurately locate the RFID tags on books. This aids in differentiating between correctly identified books and those that were misread. Similarly, Ref. [55] presents a technique that uses RFID and ML methods to precisely localize bookshelves. This method significantly enhances the precision in identifying books situated in specific rows, cabinets, and racks. Another technique, as mentioned in Refs. [56–58], relies on RSS data from RFID combined with DL to locate books on shelves.

The success of a personalized book service may be evaluated in part by how well it captures the reader's attention to a certain title. In Ref. [59], researchers combine RNN DL with CRFID to record users' book-related behaviors, including browsing, rearranging, page-turning, borrowing, and reading. Data on these activities are gathered through CRFID, which inherently has sequential characteristics, making it particularly suited for RNN-based ML. This approach aids in offering valuable recommendations to meet reader demands. The detailed process of Ref. [59] is depicted in Figure 3.12.

3.16.2 A Sustainable Smart Library with Practical Applications

When thinking about the long-term health of human civilization, the carbon footprint is a major factor to consider [60]. The day-to-day maintenance and administration of a library require a large number of resources. Moreover, they produce notable metabolic by-products,

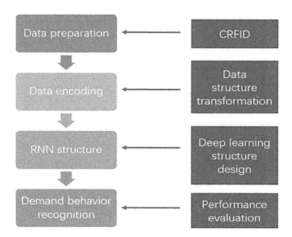

FIGURE 3.12   Intelligent book sorting system architectural diagram.

including carbon footprints. Therefore, there is an urgent need to manage resource consumption sustainably. AI-enhanced IoT can strategically schedule sustainability based on real-world requirements. As documented in Ref. [61], a series of AI-supported IoT methods have been introduced to adjust the lighting in libraries, optimizing natural light usage. The smart mode may be activated by going into the device's settings menu. The system's sensors then measure light intensity and angle, adjusting autonomously based on user settings, as depicted in Figure 3.13. In another study [62], a system designed to boost sustainability efficiency in smart libraries is introduced. Several sensor networks, servers, as well as sensors are all part of this system. These sensors gather data on aspects like temperature, humidity, and user details, as illustrated in Figure 3.14. All data are harmonized with the server, allowing the server's control

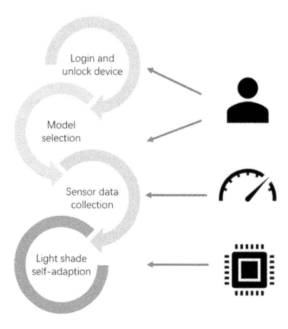

FIGURE 3.13   Depicts the smart light shade system's management structure.

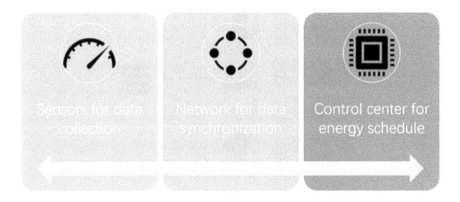

FIGURE 3.14   The smart sustainability managing framework's practical procedure for accomplishing tasks.

center to make smart decisions on equipment operation. For instance, lighting can be adjusted based on historical and real-time data. Lastly, Ref. [63] presents an innovative approach to maintaining optimal environmental conditions. This approach utilizes multiple sensors and visual data monitoring to ensure the best possible environment for preservation.

3.16.3 Applications for Secure Smart Libraries

Libraries provide a tranquil and convenient environment for readers. However, in such public settings, there's a potential risk of unintentionally exposing personal items and private data. As we usher in the era of big data and its related processing methodologies, the demand for privacy safeguards will increase. Leveraging AI-powered IoT can significantly reduce the risk associated with the exposure of personal items and privacy. Studies mentioned in Ref. [64] suggest an innovative authentication framework designed to protect reader privacy. Specifically, in Ref. [64], the data gathered through RFID are encrypted using a smart authentication algorithm, enabling secure transactions through multiple data interactions, as depicted in Figure 3.15. Using sophisticated fuzzy sets as well as case-based reasoning, the IoT risk warning system introduced in Ref. [65] can keep tabs on the health of connected devices and user behavior without any intervention.

IoT warning system with AI described in Ref. [65] has many steps in its operation:

- Risk abnormalities are detected through sensors.

- Information gathered by these sensors is sent to an IoT-based assessment and processing layer.

- Within the processing layer, this information is compared with records of unusual occurrences that have been saved over time.

- The evaluated data are then passed on to the decision-making layer, where fuzzy sets as well as case-based reasoning are used to classify the precise nature of the anomaly.

- Once a decision is reached, the historical dataset is updated and reserved for future data processing and comparison, as illustrated in Figure 3.16.

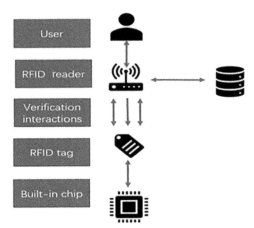

FIGURE 3.15   The structure of a smart RFID authentication system workflow.

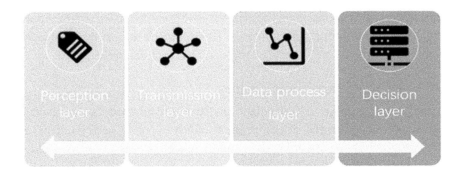

FIGURE 3.16   Illustrates the structure of an IoT warning system enhanced with AI workflow.

3.17 CONCLUSION

In conclusion, the convergence of the Internet of Things (IoT) and Artificial Intelligence (AI) marks a significant milestone in technological advancement, offering groundbreaking possibilities that promise to redefine our interaction with the physical and digital worlds. This confluence, often referred to as AIoT, amalgamates the pervasive connectivity of IoT with the cognitive capabilities of AI, paving the way for smarter devices, more intuitive interfaces, and streamlined operations.

Both IoT and AI independently have brought about significant changes, making our lives more convenient, efficient, and interconnected. When fused, they promise a world where cities are smarter, industries are more automated, and everyday experiences, from household tasks to library visits, are enhanced. The potential applications span from urban planning and traffic management to revolutionizing the library experience with AIoT-driven solutions, such as personalized book services, intelligent seat management, and efficient resource consumption.

However, the marriage of these two domains is not without its challenges. The extensive data collection and sharing brought about by IoT devices, coupled with the processing power and decision-making capabilities of AI, amplify concerns related to security, privacy, and complexity. As these technologies become more intertwined, the vulnerabilities they may introduce become more intricate, requiring a proactive approach from all stakeholders. Cybersecurity, privacy risks, and the complexities in integrating and managing AIoT technologies are issues that need earnest attention. The cost implications, while providing transformative outcomes, must also be assessed judiciously.

It is of paramount importance for manufacturers, developers, governments, and consumers to prioritize security, privacy, and ethical considerations. Only with a balanced approach can we harness the full potential of the AIoT synergy, ensuring that the benefits are maximized, and the associated risks are minimized. As we step into this new era of interconnected intelligence, the challenge lies not just in technological innovation but in responsible and informed adoption. The future beckons a world transformed by AIoT, but it is up to us to navigate this journey wisely and ethically.

REFERENCES

[1] Riahi A, Challal Y, Natalizio E, Chtourou Z, Bouabdallah A, "A systemic approach for IoT security," In *2013 IEEE International Conference on Distributed Computing in Sensor Systems*, Cambridge, MA, USA, 2013, pp. 351–355.

[2] Chan M, Campo E, Estève D, Fourniols JY, "Smart homes-current features and future perspectives," *Maturitas*, vol. 64, no. 2, pp. 90–97, 2009.

[3] Stojkoska BLR, Trivodaliev KV, "A review of Internet of Things for smart home: Challenges and solutions," *Journal of Cleaner Production*, vol. 140, pp. 1454–1464, 2017.

[4] Khan WU, Imtiaz N, Ullah I, "Joint optimization of NOMA-enabled backscatter communications for beyond 5G IoT networks," *Internet Technology Letters*, vol. 4, no. 2, p. e265, 2021.

[5] Gupta D, Juneja S, Nauman A, Hamid Y, Ullah I, Kim T, Tag eldin EM, Ghamry NA, "Energy saving implementation in hydraulic press using industrial Internet of Things (IIoT)," *Electronics*, vol. 11, no.23, p. 4061, 2022.

[6] Vermesan O, Friess P, *Internet of Things: Converging Technologies for Smart Environments and Integrated Ecosystems*, Aalborg: River Publishers, 2013.

[7] Ullah I, Qian S, Deng Z, Lee JH, "Extended Kalman filter-based localization algorithm by edge computing in wireless sensor networks," *Digital Communications and Networks*, vol. 7, no. 2, pp. 187–195, 2021.

[8] Mazhar T, Irfan HM, Haq I, Ullah I, Ashraf M, Shloul TA, Ghadi YY, Imran, Elkamchouchi DH, "Analysis of challenges and solutions of IoT in smart grids using AI and machine learning techniques: A review," *Electronics*, vol. 12, no.1, p. 242, 2023.

[9] Suciu G, Vulpe A, Halunga S, Fratu O, Todoran G, Suciu V, "Smart cities built on resilient cloud computing and secure Internet of Things," In *2013 19th International Conference on Control Systems and Computer Science*, Bucharest, Romania, May 2013, pp. 513–518.

[10] Pal R, Adhikari D, Heyat MB, Ullah I, You Z, "Yoga meets intelligent Internet of Things: Recent challenges and future directions," *Bioengineering*, vol. 10, no. 4, p. 459, 2023.

[11] Khan HU, Hussain A, Nazir S, Ali F, Khan MZ, Ullah I, "A service-efficient proxy mobile IPv6 extension for IoT domain," *Information*, vol. 14, no. 8, p. 459, 2023.

[12] Al-Fuqaha A, Guizani M, Mohammadi M, Aledhari M, Ayyash M, "Internet of Things: A survey on enabling technologies, protocols, and applications," *IEEE Communications Surveys & Tutorials*, vol. 17, no. 4, pp. 2347–2376, 2015.

[13] Alcaide A, Palomar E, Montero-Castillo J, Ribagorda A, "Anonymous authentication for privacy-preserving IoT target-driven applications," *Computers & Security*, vol. 37, pp. 111–123, 2013.

[14] Miorandi D, Sicari S, De Pellegrini F, Chlamta I, "Internet of Things: Vision, applications and research challenges," *Ad Hoc Networks*, vol. 10, no. 7, pp. 1497–1516, 2012.

[15] Abdmeziem MR, Tandjaoui D, "Tailoring mikey-ticket to e-health applications in the context of Internet of Things," In *International Conference on Advanced Networking, Distributed Systems and Applications (Short Papers)*, 2014, Bejaia, Algeria, pp. 72–77.

[16] Schurgers C, Srivastava MB, "Energy efficient routing in wireless sensor networks," *IEEE Military Communications Conference MILCOM. Communications for Network-Centric Operations: Creating the Information Force*, 200, pp. 357–361, 2001.

[17] Guoqiang H, Wee Peng T, Yonggang W, "Cloud robotics: Architecture, challenges and applications," *IEEE Network*, vol. 26, no. 3, pp. 21–28, 2012.

[18] Michalski RS, Carbonell JG, Mitchell TM, *Machine Learning: An Artificial Intelligence Approach*, Berlin, Germany: Springer Science & Business Media, 2013.

[19] Witten IH, Frank E, "Data mining: Practical machine learning tools and techniques," In Morgan Kaufmann. Amsterdam: Elsevier, 2006, 558 pages.

[20] Monostori L, Kádár B, Bauernhansl T, et al., "Cyber-physical systems in manufacturing," *CIRP Annals*, vol. 65, no. 2, pp. 621–641, 2016.

[21] Lee EA, Seshia SA, *Introduction to Embedded Systems: A Cyber-Physical Systems Approach*, Cambridge, MA: MIT Press, 2016.

[22] Hassan QF, Khan AR, Madani SA, "Internet of Things: Challenges, advances, and applications," In *Chapman & Hall/CRC Computer and Information Science Series*, Boca Raton, FL: CRC Press, 2017, 436 pages.

[23] Asif M, Khan WU, Afzal HR, Nebhen J, Ullah I, Rehman AU, Kaabar MK, "Reduced-complexity LDPC decoding for next-generation IoT networks," *Wireless Communications and Mobile Computing*, vol. 2021, pp. 1–10, 2021.

[24] Mazhar T, Talpur DB, Shloul TA, Ghadi YY, Haq I, Ullah I, Ouahada K, Hamam H, "Analysis of IoT security challenges and its solutions using artificial intelligence," *Brain Sciences*, vol. 13, no. 4, p. 683, 2023.

[25] Fortino G, Trunfio P, *Internet of Things Based on Smart Objects: Technology, Middleware and Applications*, New York: Springer, 2014.

[26] Khan HU, Sohail M, Ali F, Nazir S, Ghadi YY, Ullah I, "Prioritizing the multi-criterial features based on comparative approaches for enhancing security of IoT devices," *Physical Communication*, vol. 59, p. 102084, 2023.

[27] Ullah I, Chen J, Su X, Esposito C, Choi C, "Localization and detection of targets in underwater wireless sensor using distance and angle-based algorithms," *IEEE Access*, vol. 7, pp. 45693–45704, 2019.

[28] Yang LT, Di Martino B, Zhang Q, "Internet of everything," *Mobile Information System*, vol. 17, pp. 1–3, 2017.

[29] Baheti R, Gill H, "Cyber-physical systems," *Impact Control Technology*, vol. 12, no. 1, pp. 161–166, 2011.

[30] Gorman MM, *Database Management Systems: Understanding and Applying Database Technology*. Butterworth-Heinemann, Oxford, 2014.

[31] Theodoridis S, Koutroumbas K, *Pattern Recognition*, Elsevier Science, 2008.

[32] Marz N, Warren J, *Big Data: Principles and Best Practices of Scalable Real-Time Data Systems*, Simon and Schuster, New York, 2015.

[33] Raza MA, Abolhasan M, Lipman J, Shariati N, Ni W, Jamalipour A, "Statistical learning-based grant-free access for delay-sensitive Internet of Things applications," *IEEE Transactions on Vehicular Technology*, vol. 71, no. 5, pp. 5492–5506, 2022.

[34] Mazhar T, Irfan HM, Khan S, Haq I, Ullah I, Iqbal M, Hamam H, "Analysis of cyber security attacks and its solutions for the smart grid using machine learning and blockchain methods," *Future Internet*, vol. 15, no. 2, p. 83, 2023.

[35] Khan I, Tian YB, Ullah I, Kamal MM, Ullah H, Khan A, "Designing of E-shaped microstrip antenna using artificial neural network," *International Journal of Computing, Communication and Instrumentation Engineering*, vol. 5, no.1, pp. 23–26, 2018.

[36] Cui Q, Zhang Z, Shi Y, Ni W, Zeng M, Zhou M, "Dynamic multichannel access based on deep reinforcement learning in distributed wireless networks," *IEEE Systems Journal*, vol. 16, no. 4, pp. 5831–5834, 2021.

[37] Emami Y, Wei B, Li K, Ni W, Tovar E, "Joint communication scheduling and velocity control in multi-UAV-assisted sensor networks: A deep reinforcement learning approach," *IEEE Transactions on Vehicular Technology*, vol. 70, no. 10, pp. 10986–10998, 2021.

[38] Li K, Ni W, Dressler F, "LSTM-characterized deep reinforcement learning for continuous flight control and resource allocation in UAV-assisted sensor network," *IEEE Internet of Things Journal*, vol. 9, no. 6, pp. 4179–4189, 2021.

[39] Daniel OC, Ramsurrun V, Seeam AK, "Smart library seat, occupant and occupancy information system, using pressure and RFID sensors," In *Proceedings of the 2019 Conference on Next Generation Computing Applications (NextComp)*, Mauritius, 2019, pp. 1–5.

[40] Maepa MR, Moeti MN, "IoT-based smart library seat occupancy and reservation system using RFID and FSR technologies for South African Universities of Technology," In Proceedings of the International Conference on Artificial Intelligence and its Applications, Association for Computing Machinery, New York, 2021, pp. 1–8.

[41] Liu Y, Ye H, Sun H, "Mobile phone library service: Seat management system based on WeChat," *Library Management*, vol. 42, no. 6/7, pp. 421–435, 2021.

[42] Zhou D, "Case study on seat management of university library based on WeChat public number client-taking Jianghan University library as an example," In *Proceedings of the 2019 4th International Conference on Mechanical, Control and Computer Engineering (ICMCCE)*, Hohhot, China, 2019, pp. 630–6303.

[43] Upala M, Wong WK, "IoT solution for smart library using facial recognition," In *IOP Conference Series: Materials Science and Engineering*, vol. 495, no. 1, pp. 012030, 2019.

[44] Antevski K, Redondi AEC, Pitic R, "A hybrid BLE and Wi-Fi localization system for the creation of study groups in smart libraries," In *Proceedings of the 2016 9th IFIP Wireless and Mobile Networking Conference (WMNC)*, Colmar, France, 2016, pp. 41–48.

[45] Shi X, Tang K, Lu H, "Smart library book sorting application with intelligence computer vision technology," *Library Hi Tech*, vol. 39, no. 1, pp. 220–232, 2021.

[46] Martinez-Martin E, Ferrer E, Vasilev I, del Pobil AP, "The UJI aerial librarian robot: A quadcopter for visual library inventory and book localisation," *Sensors*, vol. 21, no. 4, p. 1079, 2021.

[47] Anoop A, Ubale NA, "Cloud based collaborative filtering algorithm for library book recommender system," In *Proceedings of the 2020 Third International Conference on Smart Systems and Inventive Technology (ICSSIT)*, Tirunelveli, India, 2020, pp. 695–703.

[48] Khalil H, Rahman SU, Ullah I, Khan I, Alghadhban AJ, Al-Adhaileh MH, Ali G, ElAffendi M, "A UAV-swarm-communication model using a machine-learning approach for search-and-rescue applications," *Drones*, vol. 6, no. 12, p. 372, 2022.

[49] Abideen ZU, Mazhar T, Razzaq A, Haq I, Ullah I, Alasmary H, Mohamed HG. "Analysis of enrollment criteria in secondary schools using machine learning and data mining approach," *Electronics*, vol. 12, no. 3, p. 694, 2023.

[50] Karthikeyan D, Arumbu VP, Surendhirababu K, Selvakumar K, Divya P, Suhasini P, Palanisamy R, "Sophisticated and modernized library running system with OCR algorithm using IoT," *Indonesian Journal of Electrical Engineering and Computer Science*, vol. 24, no. 3, pp. 1680–1691, 2021.

[51] Ahmad S, Ullah T, Ahmad I, Al-Sharabi A, Ullah K, Khan RA, Rasheed S, Ullah I, Uddin MN, Ali MS, "A novel hybrid deep learning model for metastatic cancer detection," *Computational Intelligence and Neuroscience*, vol. 2022, 2022, 14 pages.

[52] Khan S, Ullah I, Ali F, Shafiq M, Ghadi YY, Kim T, "Deep learning-based marine big data fusion for ocean environment monitoring: Towards shape optimization and salient objects detection," *Frontiers in Marine Science*, vol. 9, p. 1094915, 2023.

[53] Li J, Liu Y, Wang L, "Design and development of promotion APP of university smart library service platform based on network teaching," In *Proceedings of the 2021 Fifth International Conference on I-SMAC (IoT in Social, Mobile, Analytics and Cloud) (I-SMAC)*, Palladam, India, 2021, pp. 1344–1347.

[54] Bi S, Wang C, Wu B, Gong Y, Ni W, "An accurate book-localization approach based on passive ultra-high-frequency RFID," In *International Conference on Computing, Control and Industrial Engineering*, Springer Nature Singapore, Singapore, 2021, pp. 584–591.

[55] Yaman O, Ertam F, Tuncer T, Firat Kilincer I, "Automated UHF RFID-based book positioning and monitoring method in smart libraries," *IET Smart Cities*, vol. 2, no. 4, pp. 173–180, 2020.

[56] Cheng S, Wang S, Guan W, Xu H, Li P, "3DLRA: An RFID 3D indoor localization method based on deep learning," *Sensors*, vol. 20, no. 9, pp. 2731, 2020.

[57] Waleed S, Ullah I, Khan WU, Rehman AU, Rahman T, Li S, "Resource allocation of 5G network by exploiting particle swarm optimization," *Iran Journal of Computer Science*, vol. 4, no. 3, pp. 211–219, 2021.

[58] Rasheed Z, Ma YK, Ullah I, Al Shloul T, Tufail AB, Ghadi YY, Khan MZ, Mohamed HG, "Automated classification of brain tumors from magnetic resonance imaging using deep learning," *Brain Sciences*, vol. 13, no. 4, p. 602, 2023.

[59] Bai R, Zhao J, Li D, Lv X, Wang Q, Zhu B, "RNN-based demand awareness in smart library using CRFID, *China Communications*, vol. 17, no. 5, pp. 284–294, 2020.

[60] Bi S, Fang Z, Yuan X, Wang X, "Joint base station activation and coordinated downlink beamforming for HetNets: Efficient optimal and suboptimal algorithms," *IEEE Transactions on Vehicular Technology*, vol. 68, no. 4, pp. 3702–3712, 2019.

[61] Xue J, Wang Y, Wang M, "Smart design of portable indoor shading device for visual comfort-A case study of a college library," *Applied Sciences*, vol. 11, no. 22, p. 10644, 2021.

[62] Yang CJ, Kang HB, Zhang L, Zhang RY, "A design of smart library sustainability consumption monitoring and management system based on IoT," In *Advances in Intelligent Systems and Computing, Proceedings of the Fifth Euro-China Conference on Intelligent Data Analysis and Applications (ECC 2018)*, Springer International Publishing, Xian, China, 2019, pp. 217–224.

[63] Monti L, Mirri S, Prandi C, Salomoni P, "Preservation in smart libraries: An experiment involving IoT and indoor environmental sensing," In *Proceedings of the 2019 IEEE Global Communications Conference (GLOBECOM)*, Waikoloa, HI, 2019, pp. 1–6.

[64] Adeniji OD, Rukayat O, Solomon A, "Securing privacy risks associated with radio frequency identification based library management system," *International Journal of Academic and Applied Research (IJAAR)*, vol. 4, pp. 178–182, 2020.

[65] Xie Y, Liu J, Zhu S, Chong D, Shi H, Chen Y, "An IoT-based risk warning system for smart libraries," *Library Hi Tech*, vol. 37, no. 4, pp. 918–932, 2019.

II

The Future of Data Analytics in Communication

CHAPTER 4

The Future of Artificial Intelligence in Communication

Sadaf Hussain

Lahore Garrison University

Tanweer Sohail

University of Jhang

Rabia Afzaal

Lahore Garrison University

Muhammad Adnan Khan

Riphah International University
Skyline University College
Gachon University

4.1 INTRODUCTION: BACKGROUND AND DRIVING FORCES

Artificial intelligence (AI) has rapidly advanced and become a ubiquitous presence in our daily lives [1]. It is being used across various industries for efficiency gains and cost reductions. AI has also brought about significant changes in the way we work and live, becoming a national strategy for development and application [2]. From voice assistants like Siri and Alexa to chatbots and virtual assistants, AI is now an integral part of the way we communicate. With ChatGPT being the fastest growing "app" on record, it's clear that AI will have a huge impact on us, society, and the way we communicate. AI has found extensive applications in clinical medicine, aiding in image analysis, lesion determination, healthcare management, and disease prediction [3]. AI is rapidly evolving and is expected to have a significant impact on communication in the future [4,5].

DOI: 10.1201/9781032648309-6

AI is already being applied in various industries, including the news industry, where it is used for news gathering, content production, distribution, and consumer consumption [6]. In the field of smart grid, AI techniques are being used to address the challenges of integrating renewable energy sources into the electrical grid [7]. The development and deployment of AI have economic, social, and geopolitical implications, which require international policy coordination and cooperation [8]. AI technology is still in the research and development stage, with the potential to enhance human welfare and benefit humanity. Overall, AI is poised to reshape the way we connect and interact, offering new possibilities and opportunities for communication.

This chapter explores the exciting prospects of AI in the future of communication, delving into the transformative impact it is expected to have on various aspects of human interaction.

4.1.1 AI in Communication

One of the most significant benefits of AI in communication is its ability to improve accessibility and efficiency. For example, chatbots and virtual assistants can provide instant responses to inquiries and customer service requests, freeing up human customer service agents to focus on more complex tasks. This can lead to high availability, customer satisfaction, low response time, and cost savings [9]; they can be used in various settings such as care management for monitoring patients' health conditions and recovery, resulting in improved outcomes and strong trusting relationships with care managers [10]. Firms using virtual agents (VAs) in service encounters should maximize perceived VA humanness, as it positively affects customer satisfaction [11]. Systems and methods that manage conversations in real time with human customers based on a dynamic and unscripted conversation flow with a virtual assistant can offer significant improvements to chatbot conversational experiences, providing a more natural and efficacious dialogue experience. Overall, chatbots and virtual assistants have the potential to enhance customer service experiences and optimize resource allocation. AI-powered translation services can also break down language barriers and facilitate communication between individuals who speak different languages [12].

AI can also enhance the personalization of communication. With access to large amounts of data [13,14]. AI can also be used to create personalized symbols for individuals who struggle with speech or writing [13]. Additionally, AI can enable the personalization of wireless networks based on each user's actual quality of service requirements and context, optimizing user satisfaction levels [14]. Furthermore, AI can be used for hyper-personalization in online business, addressing users' real-time needs and delivering the right information at the right time to the right customer through the right channel [15].

AI-powered tools can analyze user behavior and preferences, tailoring communication to suit the individual's needs. Additionally, AI can use predictive capabilities to anticipate what a user may need or want, providing suggestions and recommendations that can enhance the user's experience. For example, Spotify uses AI to curate personalized playlists for its users, based on their listening history and preferences [16,17].

Another benefit of AI in communication is its ability to analyze and interpret large amounts of data, providing insights that can help individuals and organizations make more informed decisions [18]. For example, social media monitoring tools can use AI to analyze trends and sentiment across social media platforms, helping businesses to better understand their audience and improve their marketing strategies.

4.1.1.1 The Challenges and Risks of AI in Communication

The challenges and opportunities of distinguishing between AI-generated and human-generated communication are explored in the provided abstracts. One challenge is the Replicant Effect, where participants mistrust hosts whose profiles are labeled as or suspected to be written by AI. While AI has the potential to revolutionize communication, it also poses certain challenges and risks. For example, as AI becomes more sophisticated, it may become difficult to differentiate between AI-generated and human-generated communication [19]. This could lead to issues with trust and transparency, particularly in areas such as journalism and marketing. Additionally, AI algorithms may perpetuate biases and reinforce existing inequalities if they are not properly designed and tested.

Research has shown that AI language models can perpetuate gender and racial biases, reflecting the biases that are present in the data they are trained on. Studies have demonstrated that language models inherit higher bias when trained on unbalanced data, but using debiased pre-trained embeddings can help reduce bias [20]. For example, a study by the AI Now Institute found that popular language models such as GPT-2, 3, 4 and BERT have gender biases, with male pronouns and names being more frequently associated with career-related words than female pronouns and names [21]. Furthermore, research has shown that gender bias is also present in syntax textbooks and academic articles, where women are underrepresented and subjected to gender stereotypes [22]. These findings highlight the need for awareness and efforts to address gender biases in language models, academic publications, and everyday speech.

4.1.1.2 The Future of AI in Communication

Despite the challenges and risks of AI in communication, the potential benefits are too significant to ignore. AI has the ability to enhance efficiency and productivity, freeing communication professionals to focus on the creative side, strategy, and analytical thinking [23]. As AI continues to advance, it can become more sophisticated and capable of understanding and responding to human communication in more natural and nuanced ways [24]. The development of humanoid robots, such as Sophia, demonstrates the potential for AI to interact with humans in a lifelike and natural manner [25]. However, it is important to ensure that AI is developed with safety, transparency, and ethical considerations in mind. By addressing these challenges and harnessing the potential of AI, communication professionals can benefit from its capabilities while also being aware of its impacts and ensuring its responsible use [26].

Additionally, the use of AI in communication is expected to increase in fields such as healthcare, education, and customer service. For example, telemedicine and teletherapy services are using AI-powered chatbots to provide instant mental health support to patients. AI is also being used to develop personalized educational content and provide students with individualized feedback.

4.1.1.3 Enhanced Personalization

In the foreseeable future, AI is set to revolutionize communication by ushering in an era of unparalleled personalization. As AI systems gain insights from user behaviors and preferences, they will refine content delivery, recommendations, and responses to cater to individual requirements, thus crafting highly unique and captivating user experiences. However, this progression also raises essential concerns. The need for extensive user data to fuel sophisticated AI systems for personalization calls attention to privacy and security issues [27]. Moreover, the potential for AI systems to perpetuate biases present in human-generated data presents challenges to fairness and equality. Additionally, AI's struggle to comprehend and respond to human emotions remains an obstacle to effective communication. To harness AI's potential, transparency and accountability are paramount, ensuring users comprehend AI processes and data usage. The looming threat of job displacement demands equitable solutions for sharing AI benefits. Furthermore, as AI-driven communication tools proliferate, they might be exploited to disseminate misinformation, eroding public discourse and trust. The loss of control over increasingly potent AI systems poses risks to privacy and security, underscoring the need for responsible development and deployment strategies.

4.1.1.4 Natural Language Processing

A dramatic revolution in AI-driven communication is about to occur because to developments in natural language processing (NLP). The line between human and machine contact is increasingly being blurred as voice assistants and chatbots get more sophisticated and are able to comprehend context, emotions, and language nuances. The widespread deployment of AI in communication may be impacted by issues including data privacy, bias, emotional intelligence, transparency, accountability, job displacement, false information, eroding trust, and loss of control as these innovations take shape [18]. Moreover, additional concerns include the potential development of autonomous AI systems communicating independently, the misuse of AI for social engineering and criminal activities, and the ethical implications of these advancements. Taking on these complex problems is essential to responsibly use AI's potential for communication, highlighting the significance of a complete knowledge of AI's ethical implications and the implementation of safeguards to assure its beneficial effects [28,29].

4.1.1.5 Seamless Multilingual Communication

In the realm of future AI-driven communication, language barriers are rapidly dissolving thanks to advanced translation tools that offer real-time cross-cultural understanding and collaboration [30]. While difficulties like accuracy, biases, adoption barriers, privacy worries, job displacement, and potential abuse for propaganda and impersonation must carefully be addressed to ensure ethical and beneficial integration, AI language models have the potential to bridge linguistic divides as they continue to develop [31]. The integration of

AI into translation services brings forth a set of pivotal challenges. First, the advancement of accuracy and fluency is ongoing, yet concerns linger over potential misunderstandings and communication breakdowns due to lingering inaccuracies and awkward phrasing. Second, the issue of cultural sensitivity emerges as AI language models inherit biases from training data, potentially leading to unjust treatment or bias against certain groups. Third, despite its transformational potential, user apprehension and distrust pose obstacles to the acceptance and deployment of AI-powered translation. Fourth, as AI-driven translation becomes more complex, more user data must be accessed, raising issues with data security and privacy. Finally, the transformative power of AI translation also poses a dilemma of potential job displacement, demanding deliberate strategies to ensure the fair distribution of AI's advantages [32]. Beyond these, other potential challenges include the potential misuse of AI for censorship, propaganda, or criminal purposes, such as generating fake news or impersonation. The deployment of strong safeguards and a thorough knowledge of AI's ethical implications are required to address these issues and enable its appropriate and beneficial inclusion into international communication.

4.1.1.6 Empowering Remote Collaboration

AI-powered communication tools are expected to further disrupt the rapidly changing communication scene, which has already seen a growth in remote work and virtual collaboration spurred on by recent international events. Through the use of these tools, teams will be able to collaborate more effectively and productively across geographic borders by streamlining scheduling, automating administrative processes, and improving video conferencing [33]. However, this transformative potential also introduces a series of challenges. These encompass concern about data privacy and security due to the increasing sophistication of AI tools, the risk of biases from human-generated training data influencing communication, the ongoing struggle of AI to comprehend and respond to human emotions effectively, the necessity of transparency and accountability for user trust, the potential for job displacement, the misuse of AI tools for misinformation and trust erosion, and the potential loss of human control over increasingly potent AI systems. AI-powered communication tools are expected to further disrupt the rapidly changing communication scene, which has already seen a growth in remote work and virtual collaboration spurred on by recent international events. Through the use of these tools, teams will be able to collaborate more effectively and productively across geographic borders by streamlining scheduling, automating administrative processes, and improving video conferencing [34].

4.1.1.7 Predictive Insights and Decision-Making

The future of AI in communication holds the promise of insightful and strategic platforms that leverage AI's ability to provide predictive insights, enabling businesses and individuals to make informed decisions by identifying trends and patterns [35,36]. However, this evolution is accompanied by a set of challenges. These encompass concerns regarding data privacy and security, as AI-powered communication platforms require access to more user data, as well as the potential for biases derived from human-generated training data, which could lead to unjust or discriminatory outcomes. Emotional intelligence remains a hurdle, as AI platforms struggle to fully grasp and respond to human emotions, impacting the nuanced nature of communication. The potential for AI-powered platforms to spread misinformation, erode trust, and disrupt human control raises critical concerns. To guarantee that AI is responsibly included in the future of communication, these problems should be addressed and reduced by thorough ethical considerations and strong protections [37,38]. As AI's capacity to analyze enormous volumes of data in real-time grows, it creates opportunities for more strategic and intelligent communication. AI has the potential to customize communications, optimize marketing campaigns, uncover consumer behavior trends, and anticipate attrition. To ensure ethical and secure use, it is crucial to remember that AI is still a new technology, therefore these improvements must be treated with careful consideration of problems.

4.1.1.8 Ethical Considerations

In the evolving landscape of communication, the increasing prominence of AI brings ethical considerations to the forefront. As AI gains a central role, safeguarding data privacy, rectifying biases in algorithms, and upholding transparency are pivotal to building trust and promoting responsible AI integration into communication practices [39]. AI-powered communication systems, as they become more sophisticated, necessitate access to greater user data, raising legitimate concerns about data privacy and security. Instilling user confidence will depend on ensuring appropriate data usage and protecting privacy. Furthermore, the possibility of biases in AI systems originating from human-generated data highlights the significance of creating impartial AI systems that treat all users fairly. Transparency and accountability must prevail, allowing users to comprehend how AI systems operate and how their data are utilized, granting them the ability to control and limit data usage as desired. Trust in AI communication systems hinges on developers' transparency regarding system functioning and data usage, fostering widespread and responsible adoption [40–42]. Embracing responsible AI deployment demands not only societal benefits but also careful consideration of potential repercussions, such as misinformation dissemination or erosion of privacy. Addressing these challenges is imperative to ensure that AI contributes positively and ethically to society's communication landscape [43].

4.2 LITERATURE SURVEY

The importance of communication is unmatched in today's digital environment because of technological breakthroughs. The development of technology and the internet has changed how we communicate, work together, and share information. Today, effective communication is essential for success because it enables people to take advantage of a variety of technological options. The capacity to interact with people worldwide, exchange ideas through blogs and podcasts, develop personal connections, work together easily, and stay on top of trends has made communication the key to using the digital landscape.

The integration of AI has further propelled the evolution of communication [18]. AI has simplified, automated, and increased the efficiency of communication thanks to advancements in NLP and machine learning. With the ability to utilize vocal commands, virtual assistants like Siri, Alexa, and Google Assistant have changed how people engage with technology [44]. AI makes communication more accessible, benefiting those with disabilities through text-to-speech and speech-to-text technologies [45]. In the realm of AI-driven innovation, ONPASSIVE stands out as a trailblazer. AI-based companies pioneer cutting-edge solutions, striving to provide fully autonomous products to a global clientele [46].

Researchers have studied how AI is changing conventional communication paradigms. The way users engage with technology is changing thanks to chatbots, virtual assistants, and AI-driven recommendation systems. By automating customer service and increasing user engagement, chatbots powered by AI enable real-time conversations [47]. Virtual assistants like Siri and Google Assistant have revolutionized device-human interactions, making tasks hands-free and intuitive [48].

Personalized communication experiences are now possible because of AI's capacity to comprehend user preferences, actions, and language subtleties. User data are analyzed by AI algorithms to give customized content, increasing engagement and pleasure. User engagement and conversion rates have grown as a result of personalized marketing techniques that are fueled by AI analytics [49].

Language boundaries have been removed by AI-driven language translation systems, allowing for seamless communication across borders. With the use of real-time translation services, people may engage with others throughout the world more effectively by communicating in their native tongue [50]. AI-powered translation tools are gaining traction in fields like diplomacy, international business, and cross-cultural collaboration.

AI's introduction into communication raises several moral questions. Concerns concerning fairness and equitable representation have been highlighted as a result of bias in AI models caused by skewed training data [51]. Researchers emphasize the need for ethical AI development to mitigate biases and ensure unbiased communication outcomes.

With AI's growing dependence on user data, safeguarding privacy and security has become paramount. Ensuring user trust by transparently handling data and implementing stringent security measures is essential [52]. AI-driven encryption techniques and privacy-preserving protocols are emerging to protect sensitive communication data.

AI's integration in communication extends to enhancing collaborative and multimodal interactions. Collaborative platforms empowered by AI enable effective teamwork, project management, and coordination across geographies [53]. The incorporation of AI in interactive visualization tools offers new avenues for data-driven communication, aiding decision-making and information dissemination.

The literature emphasizes how AI is used in a variety of fields, including healthcare, finance, marketing, and education. Virtual financial advisers give individualized financial advice, medical chatbots powered by AI in the healthcare industry simplify patient contacts, and AI-driven educational platforms provide individualized learning opportunities [54,55].

Future study focuses on improving AI's comprehension of sarcasm in communication, cultural settings, and human emotions as it develops [56]. The ethical dimensions of AI in communication, along with addressing privacy concerns and ensuring transparency, are areas that require continuous exploration [57].

4.3 THE FUTURE OF AI IN DIVERSE TYPES OF COMMUNICATIONS

4.3.1 Fundamental Communication

Future study focuses on improving AI's comprehension of human emotion as it develops. There is a significant transition taking place across several disciplines with regard to the use of AI in communication in the future. A major move toward AI systems that not only comprehend and respond to human language but also do so with naturalness and empathy, cultural contexts, and sarcasm in communication is represented by improved communication [58]. This entails AI-driven communication systems seamlessly blending technological prowess with human-like interactions, resulting in more intuitive user experiences. In order to remove obstacles and make interactions with AI systems more user-friendly and emotionally engaging, it is imperative to reach this degree of naturalness. Additionally, privacy and security take on a crucial role in this growth, needing careful design to protect user data and build trust. Transparency and accountability are equally essential, as users need to have insights into how AI systems operate and utilize their data.

AI's influence on human-computer interaction (HCI) is becoming more and clearer [59]. Customer service is being redefined by chatbots, which offer quick and helpful support. By supporting users with tasks and information retrieval, virtual assistants are increasing productivity. Machine learning and language processing are advancing the way we use technology and comprehend massive volumes of data. Machine-to-Machine (M2M) communication adds another layer of complexity to the AI landscape [60]. Here, challenges include ensuring data privacy, making AI systems interpretable, preventing biases, fortifying their robustness against potential attacks, and optimizing their efficiency. These challenges are especially crucial in smart domains like transportation, manufacturing, healthcare, and cities, where AI's transformative impact is most pronounced.

As AI technology advances, its promise in M2M communication intensifies, reshaping industries and societies at large. The synergy between AI and communication promises a dynamic and evolving landscape driven by transformative technology and human ingenuity. In essence, the future of AI in communication holds the potential to foster more natural, empathetic, creative, and secure interactions while addressing the multifaceted challenges associated with responsible innovation in this field.

4.3.2 Enhanced Communication

In the rapidly evolving landscape of AI-powered communication, the challenge of achieving explainability and interpretability is becoming increasingly pivotal. It is essential for AI systems to provide clear explanations for their decisions, especially in sectors like healthcare and finance, where trust and transparency are non-negotiable. Similarly, the ability to interpret the inner workings of AI systems is crucial for debugging, troubleshooting, and eliminating biases, ensuring their accountability. However, as AI systems advance in complexity, explaining and interpreting their actions becomes more challenging, making ethical and responsible usage imperative. The absence of standardized guidelines for achieving these goals makes it difficult to compare and ensure user satisfaction. Furthermore, the high costs associated with developing and maintaining transparent AI systems can limit their accessibility to a broader audience. Building and maintaining user trust is paramount, as transparency in decision-making and the absence of biases are critical for retaining user confidence in AI systems [61]. To navigate these intricate challenges in communication, fostering innovation and addressing ethical concerns are imperative. Current strategies to enhance explainability and interpretability encompass Explainable AI (XAI), interpretable machine learning, visualization techniques, natural language explanations, and human-in-the-loop approaches. As AI continues to advance, the evolution of strategies to tackle these challenges will continue to shape the future of AI-powered communication [62].

The efficient processing of feedback appears as a crucial obstacle for future progress in the field of AI-driven communication. AI systems when required to continuously learn and improve, user feedback must be understood and interpreted appropriately. Users must receive feedback quickly in order to learn from mistakes and improve their communication abilities. Personalized criticism is equally important since it allows for efficient progress by adapting advice to specific requirements. Scalability is crucial because AI systems need to effectively handle massive amounts of feedback for continuous improvement [63].

Security is still of utmost importance, protecting user privacy and preventing any unauthorized use of feedback data. To avoid bias and discrimination in feedback, it's critical to uphold ethical standards. These difficulties span the development of AI in communication, highlighting the demand for prompt action to enable AI systems for iterative learning and development. At the moment, AI is being used to improve feedback and learning using machine learning, natural language processing, reinforcement learning, and active learning approaches. Privacy-preserving safeguards are being incorporated to assure user data safety. Innovative solutions will continue to influence how these difficulties are faced and overcome as AI develops [64].

4.3.3 Specialized Communication

In the rapidly evolving landscape of AI-driven communication, specialized domains introduce a host of unique challenges that demand careful consideration for the future of technological advancements in this field. One of the primary hurdles is the acquisition of domain-specific data, which is often characterized by its scarcity and the exorbitant costs associated with its collection, especially in the case of rare or niche subjects. The development of complex specialized communication models further compounds these challenges, as they necessitate substantial data and computational resources for effective training, potentially introducing complexities in both their development and deployment [64].

Furthermore, the problem of model interpretability still warrants serious consideration since complex models can conceal the decision-making procedures behind their results, undermining user confidence and comprehension. Another crucial factor is the rising threat of bias in AI models. Such algorithms may unintentionally reinforce biases found in their training data, raising concerns about communication fairness and equity [65].

Another significant problem is security weaknesses. AI-driven communication is vulnerable to assaults because of these flaws, which seriously jeopardize both user privacy and the integrity of the communication model. The creation and application of specialized communication models must be carefully thought out in light of ethical issues, which also play a crucial role [66,67]. This includes addressing the potential for misuse and discrimination, which could have far-reaching societal implications. Current efforts are actively addressing these multifaceted challenges. Researchers are working on more efficient data collection techniques, striving to enhance model interpretability, and implementing bias mitigation strategies [68]. Additionally, security measures are being fortified to safeguard both the privacy of users and the integrity of AI systems. Ethical guidelines are being established to ensure responsible and equitable deployment.

As AI continues to evolve, it is anticipated that innovative solutions will emerge, reshaping the landscape of specialized communication. These solutions will not only tackle the complexities outlined above but also enable complexity, meta-communication stands essential, demanding AI's finesse in deciphering context and intentions. Privacy-preserving techniques ensure data security, while collaborative communication relies on AI's orchestration of tasks and conflict resolution.

Currently, AI showcases its potential in virtual reality experiences, art generation, and enhanced understanding of human dialogues. Yet, the true crescendo lies ahead. As AI evolves, it's set to redefine communication, harmonizing technology and human creativity to unlock experiential, generative, meta, secure, and collaborative communication. This impending revolution promises to reshape how we connect, transcending the boundaries of imagination [69,70].

4.3.4 Comprehensive Communication

In envisioning the future of AI in communication, a dynamic landscape unfolds, characterized by the convergence of multi-modal communication, semantic understanding, and interactive visualization. This amalgamation presents a profound shift in how humans and machines engage. The cornerstone of this development is multi-modal communication, in which AI not only understands spoken and written words but also gestures, facial expressions, and even emotional clues, enabling a richer and more complex relationship [71]. AI can now grasp context, decode meaning, and promote cogent conversations across languages and cultures thanks to the synergy of semantic understanding, breaking down barriers to communication. Importantly, interactive visualization provides a visual layer to communication, allowing complicated facts and ideas to be presented through interesting images, improving comprehension and engagement. This integrated paradigm ushers in a revolutionary era when AI-driven communication breaks through conventional barriers to produce a rich, natural, and profoundly human relationship that enhances the fundamental nature of communication itself [72,73].

A future in which AI-powered communication systems are characterized by ethical responsibility, security, and user trust.

4.3.5 Advanced Communication

As we peer into the future of AI-driven communication, a landscape of challenges and opportunities emerges, particularly in advanced communication domains. Experiential communication beckons, where AI must seamlessly weave immersive realities through virtual and augmented technologies. Generative communication is poised for transformation as AI becomes a creative muse, giving birth to novel artistic expressions [69,70].

This hierarchy classifies the many forms of AI communication according to their core characteristics, niche uses, and cutting-edge capabilities. Remember that these classifications are not absolute and that there may be some overlap. New forms of communication might also develop as AI technology advances, significantly enhancing how it interacts with people, other AI systems, and the environment.

4.4 DISCUSSION

The study under consideration examines the evolving field of AI-powered communication, highlighting both its problems and opportunities. In order to enable genuine human-machine interactions, it emphasizes the significance of naturalness, empathy, creativity, privacy, security, and responsibility in AI systems. While acknowledging the importance of these characteristics, the study falls short in offering a thorough examination of the technical difficulties at play. Additionally, it emphasizes the difficulties associated with machine-to-machine communication, with particular emphasis on data privacy, interpretability, and bias avoidance, but it might benefit from providing more specific solutions and examples from actual applications. The study emphasizes the necessity for explainability and interpretability in AI systems, although it might go further into the state of research and development at this time. Additionally, it recognizes the difficulty of efficiently managing user comments, but it lacks specific details on workable solutions. The study focuses on data gathering and bias concerns when considering specialized communication, but it might include more vivid case cases. Providing examples from the actual world and taking into account potential limits might be helpful when presenting advanced communication and complete communication paradigms. However, the article offers a useful starting point for additional investigation and debate in this developing area of AI-powered communication.

4.5 CONCLUSION

AI's incorporation into communication has developed from a purely theoretical idea to an irrefutable reality that is changing the way we connect. This revolutionary change opens up a wide range of potential outcomes, from the enhancement of customization and the removal of language barriers to the facilitation of remote cooperation and the foresight of forthcoming communication trends. This vast potential of AI calls for a careful examination of its ethical implications since it promises to take communication to previously unimaginable heights. We begin on a journey that takes communication to new heights, enabling connection and creativity that were previously restricted to the realms of fantasy by embracing AI's power for transformation while preserving ethical issues. In conclusion, understanding how AI may change communication is still in its infancy. We can change accessibility and efficiency, enhance personalization, and use predictive insights by utilizing AI's capabilities. However, just like with the introduction of any innovative technology, it is crucial to be aware of the dangers and difficulties that come with integrating AI. We may work together to steer toward a future in which AI-driven communication serves as a testimony to both efficacy and ethics by deliberately addressing these concerns.

These are only a few of the issues that must be resolved as AI advances and permeates more aspects of our daily life. To maximize the advantages of this technology while lowering the hazards, it is crucial to take a careful and moral approach to its development and usage in communication.

REFERENCES

[1] H. Hassani, E. S. Silva, S. Unger, M. TajMazinani, and S. Mac Feely, "Artificial intelligence (AI) or intelligence augmentation (IA): What is the future?," *Artif. Intell.*, vol. 1, no. 2, pp. 143–155, 2020, doi:10.3390/AI1020008.

[2] Z. Shao et al., "AI 2000: A decade of artificial intelligence," In *WebSci '20: 12th ACM Conference on Web Science*, Southampton, UK, pp. 345–354, 2020, doi:10.1145/3394231.3397925.

[3] C. Liu, D. Jiao, and Z. Liu, "Artificial intelligence (AI)-aided disease prediction," *BIO Integr.*, vol. 1, no. 3, pp. 130–136, 2020, doi:10.15212/BIOI-2020-0017.

[4] Y. Feng and X. Lv(U), "Frontier application and development trend of artificial intelligence in new media in the AI era," *Lect. Notes Data Eng. Commun. Technol.*, vol. 97, pp. 58–64, 2022, doi:10.1007/978-3-030-89508-2_8/COVER.

[5] M. Massaoudi, S. S. Refaat, and H. Abu-Rub, "On the pivotal role of artificial intelligence toward the evolution of smart grids," *Smart Grid Enabling Technol.*, pp. 359–420, 2021, doi:10.1002/9781119422464.CH15.

[6] C. Feijóo et al., "Harnessing artificial intelligence (AI) to increase wellbeing for all: The case for a new technology diplomacy," *Telecomm. Policy*, vol. 44, no. 6, p. 101988, 2020, doi:10.1016/J.TELPOL.2020.101988.

[7] A. L. Guzman and S. C. Lewis, "Artificial intelligence and communication: A human-machine communication research agenda," *New Media & Society*, vol. 22, no. 1, pp. 70–86, 2019, doi:10.1177/1461444819858691.

[8] J. Tong, "Research on the influence of the new media era on the rise and development of artificial intelligence technology," In *6th Asia-Pacific Conference on Social Sciences, Humanities (APSSH 2019)*, Manila, Philippines, pp. 580–584, 2019, doi:10.25236/apssh.2019.115.

[9] M. E. Schario, C. A. Bahner, T. V. Widenhofer, J. I. Rajaballey, and E. J. Thatcher, "Chatbot-assisted care management," *Prof. Case Manag.*, vol. 27, no. 1, pp. 19–25, 2022, doi:10.1097/NCM.0000000000000504.

[10] M. Söderlund and E. L. Oikarinen, "Service encounters with virtual agents: An examination of perceived humanness as a source of customer satisfaction," *Eur. J. Mark.*, vol. 55, no. 13, pp. 94–121, 2021, doi:10.1108/EJM-09-2019-0748/FULL/HTML.

[11] M. Ashfaq, J. Yun, S. Yu, and S. M. C. Loureiro, "I, Chatbot: Modeling the determinants of users' satisfaction and continuance intention of AI-powered service agents," *Telemat. Informatics*, vol. 54, p. 101473, 2020, doi:10.1016/J.TELE.2020.101473.

[12] P. Singh, A. Verma, S. Vij, and J. Thakur. "Implications & Impact of Artificial Intelligence in Digital Media: With Special Focus on Social Media Marketing". In *E3S Web of Conferences* (Vol. 399, p. 7006). EDP Sciences, 2023.

[13] M. Wald, "AI data-driven personalisation and disability inclusion," *Front. Artif. Intell.*, vol. 3, pp. 571955–571955, 2021, doi:10.3389/FRAI.2020.571955/PDF.

[14] R. Alkurd, I. Abualhaol, and H. Yanikomeroglu, "Big-data-driven and AI-based framework to enable personalization in wireless networks," *IEEE Commun. Mag.*, vol. 58, no. 3, pp. 18–24, 2020, doi:10.1109/MCOM.001.1900533.

[15] O. Soffer, "Algorithmic personalization and the two-step flow of communication," *Commun. Theory*, vol. 31, no. 3, pp. 297–315, 2021, doi:10.1093/CT/QTZ008.

[16] P. Álvarez, J. García de Quirós, and S. Baldassarri, "A web system based on spotify for the automatic generation of affective playlists," *Commun. Comput. Inf. Sci.*, vol. 1291 CCIS, pp. 124–137, 2020, doi:10.1007/978-3-030-61218-4_9.

[17] N. Dauban, C. Senac, J. Pinquier, and P. Gaillard, "Towards a content-based prediction of personalized musical preferences using transfer learning," *Proc. - Int. Work. Content-Based Multimed. Index.*, vol. 2021, pp. 1–6, 2021, doi:10.1109/CBMI50038.2021.9461911.

[18] J. Hohenstein et al., "Artificial intelligence in communication impacts language and social relationships," *Sci. Rep.*, vol. 13, no. 1, pp. 1–9, 2023, doi:10.1038/s41598-023-30938-9.

[19] S. Natale, "Communicating through or communicating with: Approaching artificial intelligence from a communication and media studies perspective," *Commun. Theory*, vol. 31, no. 4, pp. 905–910, 2021, doi:10.1093/CT/QTAA022.

[20] P. P. Liang, C. Wu, L. P. Morency, and R. Salakhutdinov, "Towards understanding and mitigating social biases in language models," *Proc. Mach. Learn. Res.*, vol. 139, pp. 6565–6576, 2021, https://arxiv.org/abs/2106.13219v1.

[21] L. Kim, D. S. Smith, B. Hofstra, and D. A. McFarland, "Gendered knowledge in fields and academic careers," *Res. Policy*, vol. 51, no. 1, p. 104411, 2022, doi:10.1016/j.respol.2021.104411.

[22] M. O. R. Prates, P. H. Avelar, and L. C. Lamb, "Assessing gender bias in machine translation – A case study with google translate," *Neural Comput. Appl.*, vol. 32, no. 10, pp. 6363–6381, 2018, doi:10.1007/s00521-019-04144-6.

[23] E. A. López Jiménez and T. Ouariachi, "An exploration of the impact of artificial intelligence (AI) and automation for communication professionals," *J. Inform. Commun. Ethics Soc.*, vol. 19, no. 2, pp. 249–267, 2020, doi:10.1108/JICES-03-2020-0034/FULL/HTML.

[24] A. Holzinger, E. Weippl, A. M. Tjoa, and P. Kieseberg, "Digital transformation for sustainable development goals (SDGs) – A security, safety and privacy perspective on AI," *Lect. Notes Comput. Sci. (including Subser. Lect. Notes Artif. Intell. Lect. Notes Bioinformatics)*, vol. 12844 LNCS, pp. 1–20, 2021, doi:10.1007/978-3-030-84060-0_1.

[25] H. Abdalkareem Mardan and S. Kakil Ahmed, "Using AI in wireless communication system for resource management and optimisation," *Period. Eng. Nat. Sci.*, vol. 8, no. 4, pp. 2068–2074, 2020, doi:10.21533/PEN.V8I4.1677.G689.

[26] M. Szollosy, "Shifting the goalposts: Reconceptualizing robots, AI, and humans," In *Minding the Future*, B. W. S. A. T. Dainton, Ed. Liverpool: Springer, 2021, pp. 219–242. doi:10.1007/978-3-030-64269-3_11.

[27] M. Liao and S. S. Sundar, "How should AI systems talk to users when collecting their personal information? Effects of role framing and self-referencing on human-ai interaction," In *CHI '21: Proceedings of the 2021 CHI Conference on Human Factors in Computing Systems*, Yokohama, Japan, May 2021, doi:10.1145/3411764.3445415.

[28] E. Aizenberg and J. van den Hoven, "Designing for human rights in AI," *arXiv Comput. Soc.*, vol. 7, no. 2, 2020, doi:10.1177/2053951720949566.

[29] L. Crompton, "The decision-point-dilemma: Yet another problem of responsibility in human-AI interaction," *J. Responsible Technol.*, vol. 7–8, p. 100013, 2021, doi:10.1016/J.JRT.2021.100013.

[30] E. Goldenthal, J. Park, S. X. Liu, H. Mieczkowski, and J. T. Hancock, "Not all AI are equal: Exploring the accessibility of AI-mediated communication technology," *Comput. Human Behav.*, vol. 125, p. 106975, 2021, doi:10.1016/j.chb.2021.106975.

[31] G. I. Winata, A. Madotto, Z. Lin, R. Liu, J. Yosinski, and P. Fung, "Language models are few-shot multilingual learners," In *Proceedings of the 1st Workshop on Multilingual Representation Learning*, Punta Cana, Dominican Republic, pp. 1–15, 2021, doi:10.18653/v1/2021.mrl-1.1.

[32] G. G. Palmer, "AI ethics: Four key considerations for a globally secure future," In *Artificial Intelligence and Global Security*, Y. R. Masakowski, Ed. Bingley: Emerald Publishing Limited, 2020, pp. 167–176. doi:10.1108/978-1-78973-811-720201010.

[33] E. Johnson, "Just @Me: Digitally-mediated team communication in a pandemic," *The 39th ACM International Conference on Design of Communication, SIGDOC 2021*, Virtual Event, pp. 315–318, 2021, doi:10.1145/3472714.3473658.

[34] W. A. Adah, N. A. Ikumapayi, and H. B. Muhammed, "The ethical implications of advanced artificial general intelligence: Ensuring responsible AI development and deployment," *SSRN Electron. J.*, 2023, 11 pages. doi:10.2139/SSRN.4457301.

[35] J. Naidoo and R. E. Dulek, "Artificial intelligence in business communication: A snapshot," *Int. J. Bus. Commun.*, vol. 59, no. 1, pp. 126–147, J2022, doi:10.1177/2329488418819139.

[36] A. Zerfass, J. Hagelstein, and R. Tench, "Artificial intelligence in communication management: A cross-national study on adoption and knowledge, impact, challenges and risks," *J. Commun. Manag.*, vol. 24, no. 4, pp. 377–389, 2020, doi:10.1108/JCOM-10-2019-0137/FULL/HTML.

[37] Y. T. Lin, T. W. Hung, and L. T. L. Huang, "Engineering equity: How AI can help reduce the harm of implicit bias," *Philos. Technol.*, vol. 34, no. 1, pp. 65–90, 2021, doi:10.1007/S13347-020-00406-7/TABLES/2.

[38] P. Moradi and K. Levy, *The Future of Work in the Age of AI*. Cornell University, New York, 2020. doi:10.1093/oxfordhb/9780190067397.013.17.

[39] H. Aldboush and M. Ferdous, "Building trust in fintech: An analysis of ethical and privacy considerations in the intersection of big data, AI, and customer trust," *Int. J. Financ. Stud.*, vol. 11, no. 3, p. 90, 2023, doi:10.3390/IJFS11030090.

[40] B. Li et al., "Trustworthy AI: From principles to practices," *ACM Comput. Surv.*, vol. 55, no. 9, 2023, p. 90, doi:10.1145/3555803.

[41] J. Chen, J. P. Morgan, V. Storchan, and E. Kurshan, "Beyond Fairness Metrics: Roadblocks and Challenges for Ethical AI in Practice," 2021, Accessed: September 12, 2023. [Online]. Available: https://arxiv.org/abs/2108.06217v1.

[42] B. Shneiderma. "Human-centered AI". *Issues in Science and Technology*, vol. 37, no. 2, pp. 56–61, 2021.

[43] S. C. Slota et al., "Many hands make many fingers to point: challenges in creating accountable AI," *AI Soc.*, vol. 38, pp. 1–13, 2021, doi:10.1007/S00146-021-01302-0.

[44] B. Savic, M. Milic, and S. Vlajic, "Analysis and development of the model for Google Assistant and Amazon Alexa Voice Assistants integration," In *2023 27th International Conference on Information Technology (IT)*, Zabljak, Montenegro, pp. 1–4, 2023, doi:10.1109/IT57431.2023.10078705.

[45] J. F. de Almeida, W. Gottardi, and C. H. S. Tumolo, "Automatic speech recognition and text-to-speech technologies for L2 pronunciation improvement: Reflections on their affordances," *Texto Livre*, vol. 15, p. e36736, 2022, doi:10.35699/1983-3652.2022.36736.

[46] "Top Artificial Intelligence Companies in India 2019 | GoodFirms," Accessed September 3, 2023. Available: https://innohealthmagazine.com/2020/innovation/top-artificial-intelligence-organizations-in-healthcare-in-india/%0Ahttps://www.goodfirms.co/artificial-intelligence/india%0Ahttps://innohealthmagazine.com/2020/innovation/top-artificial-intelligence-organiza.

[47] S. I. Malik, M. W. Ashfque, R. M. Tawafak, G. Al-Farsi, N. A. Usmani, and B. H. Khudayer, "A chatbot to facilitate student learning in a programming 1 course: A gendered analysis," *Int. J. Virtual Pers. Learn. Environ.*, vol. 12, no. 1, pp. 1–20, 2022, doi:10.4018/IJVPLE.310007.

[48] G. Caldarini, S. Jaf, and K. McGarry, "A literature survey of recent advances in chatbots," *Information*, vol. 13, no. 1, p. 41, 2022, doi:10.3390/INFO13010041.

[49] S. Chandra, S. Verma, W. M. Lim, S. Kumar, and N. Donthu, "Personalization in personalized marketing: Trends and ways forward," *Psychol. Mark.*, vol. 39, no. 8, pp. 1529–1562, 2022, doi:10.1002/MAR.21670.

[50] D. Moroni and M. A. Pascali, "Learning topology: Bridging computational topology and machine learning," *Pattern Recognit. Image Anal.*, vol. 31, no. 3, pp. 443–453, 2021, doi:10.1134/S1054661821030184/FIGURES/3.

[51] N. Kordzadeh and M. Ghasemaghaei, "Algorithmic bias: Review, synthesis, and future research directions," *Eur. J. Inf. Syst.*, vol. 31, no. 3, pp. 388–409, 2022, doi:10.1080/09600 85X.2021.1927212.

[52] A. Fathalizadeh, V. Moghtadaiee, and M. Alishahi, "On the privacy protection of indoor location dataset using anonymization," *Comput. Secur.*, vol. 117, 2022, p. 102665. doi:10.1016/J.COSE.2022.102665.

[53] J. S. Dhatterwal, K. S. Kaswan, A. Baliyan, and V. Jain, "Integration of cloud and IoT for smart e-healthcare," *Stud. Comput. Intell.*, vol. 1021, pp. 1–31, 2022, doi:10.1007/978-3-030-97929-4_1.

[54] M. Poser, G. C. Küstermann, N. Tavanapour, and E. A. C. Bittner, "Design and evaluation of a conversational agent for facilitating idea generation in organizational innovation processes," *Inf. Syst. Front.*, vol. 24, no. 3, pp. 771–796, 2022, doi:10.1007/S10796-022-10265-6.

[55] M. Krishna Enduri et al., "Comparative study on sentimental analysis using machine learning techniques," *Mehran Univ. Res. J. Eng. Technol.*, vol. 42, no. 1, p. 207, 2023, doi:10.22581/MUET1982.2301.19.

[56] E. Yücel Karamustafa and B. Arsan, "The future of artificial intelligence: How will emotions affect AI?," *J. Bus. Digit. Age*, vol. 5, no. 1, pp. 58–64, 2022, doi:10.46238/JOBDA.1070090.

[57] D. García-Gasulla, A. Cortés, S. Álvarez-Napagao, and U. Cortés, "Signs for ethical AI: A route towards transparency," *arXiv preprint arXiv:2009.13871*, 2020.

[58] R. Girju, "Adaptive Multimodal and Multisensory Empathic Technologies for Enhanced Human Communication," 2021, Accessed: September 12, 2023. [Online]. Available: https://www.hkuriich.org.

[59] M. Antona, G. Margetis, S. Ntoa, and H. Degen, "Special issue on AI in HCI," *Int. J. Hum. Comput. Interact.*, vol. 39, no. 9, pp. 1723–1726, 2023, doi:10.1080/10447318.2023.2177421.

[60] M. Mozgovoy and C. Suero Montero, "Special issue on machine learning and natural language processing," *Appl. Sci.*, vol. 12, no. 17, pp. 8894–8894, 2022, doi:10.3390/APP12178894.

[61] V. F. de Santana, L. M. D. F. Galeno, E. V. Brazil, A. Heching, and R. Cerqueira, "Retrospective End-User Walkthrough: A Method for Assessing How People Combine Multiple AI Models in Decision-Making Systems," 2023, Accessed: September 12, 2023. [Online]. Available: https://arxiv.org/abs/2305.07530v1.

[62] F. Sovrano, F. Vitali, and M. Palmirani, "Making things explainable vs explaining: Requirements and challenges under the GDPR," *Lect. Notes Comput. Sci. (including Subser. Lect. Notes Artif. Intell. Lect. Notes Bioinformatics)*, vol. 13048 LNAI, pp. 169–182, 2021, doi:10.1007/978-3-030-89811-3_12.

[63] K. Lee, L. Smith, and P. Abbeel, "PEBBLE: Feedback-efficient interactive reinforcement learning via relabeling experience and unsupervised pre-training," *Proc. Mach. Learn. Res.*, vol. 139, pp. 6152–6163, 2021.

[64] E. Panadero and A. A. Lipnevich, "A review of feedback models and typologies: Towards an integrative model of feedback elements," *Educ. Res. Rev.*, vol. 35, p. 100416, 2022, doi:10.1016/J.EDUREV.2021.100416.

[65] N. Picchiotti and M. Gori, "Logic constraints to feature importance," *Lect. Notes Comput. Sci. (including Subser. Lect. Notes Artif. Intell. Lect. Notes Bioinformatics)*, vol. 13196 LNAI, pp. 389–402, 2022, doi:10.1007/978-3-031-08421-8_27/COVER.

[66] Z. K. Younis and B. Mahmood, "Towards the impact of security vunnerabilities in software design: A complex network-based approach," In *6th International Engineering Conference "Sustainable Technology and Development" IEC 2020*, Erbil, Iraq, pp. 157–162, 2020, doi:10.1109/IEC49899.2020.9122923.

[67] T. S. Ustun and S. M. S. Hussain, "A review of cybersecurity issues in smartgrid communication networks," In *IEEE International Conference on Power Electronics, Control and Automation (ICPECA 2019)*, New Delhi, India, vol. 2019, 2019, doi:10.1109/ICPECA47973.2019.8975629.

[68] M. Bartolo, T. Thrush, R. Jia, S. Riedel, P. Stenetorp, and D. Kiela, "Improving question answering model robustness with synthetic adversarial data generation," In *Proceedings of the 2021 Conference on Empirical Methods in Natural Language Processing*, pp. 8830–8848, 2021, doi:10.18653/v1/2021.emnlp-main.696.

[69] M. Kaczmarek-Śliwińska, "Organisational communication in the age of artificial intelligence development. opportunities and threats," *Soc. Commun.*, vol. 5, no. 2, pp. 62–68, 2019, doi:10.2478/SC-2019-0010.

[70] S. Patel et al., "Interpretation of emergent communication in heterogeneous collaborative embodied agents," *2021 IEEE/CVF International Conference on Computer Vision (ICCV)*, Montreal, QC, Canada, 2021, pp. 15933–15943, doi: 10.1109/ICCV48922.2021.01565.

[71] J. Knight, M. Dooly, and E. Barberà, "Navigating a multimodal ensemble: Learners mediating verbal and non-verbal turns in online interaction tasks," *ReCALL*, vol. 32, no. 1, pp. 25–46, 2020, doi:10.1017/S0958344019000132.

[72] Á. Vizoso, "Information visualization and usability: Tools for human comprehension," *Stud. Big Data*, vol. 70, pp. 85–98, 2020, doi:10.1007/978-3-030-36315-4_7.

[73] I. Chowdhury, A. Moeid, E. Hoque, M. A. Kabir, M. S. Hossain, and M. M. Islam, "Designing and evaluating multimodal interactions for facilitating visual analysis with dashboards," *IEEE Access*, vol. 9, pp. 60–71, 2021, doi:10.1109/ACCESS.2020.3046623.

CHAPTER 5

Data Science Meets Intelligent Internet of Things

Inam Ullah

Gachon University

Ijaz Ahamd

University of Chinese Academy of Sciences

Muhammad Shahid Anwar

Gachon University

Yuning Tao

South China University of Technology

Muhammad Shafiq

Guangzhou University

5.1 INTRODUCTION

The fusion of data science and the Internet of Things (IoT) has heralded a new age of technical innovation and intelligent decision-making [1]. The IoT is a networked system of different computing devices and digital machineries that digitalize the physical world. IoT has already had an impact on people's lives in areas such as housing, transportation, health, food, clothes, and remote controlling. Numerous home applications can be monitored by mobile and voice commands. Various programs help users advance their life standard and even make it easier for the old and crippled to live. According to MGI's report, IoT will generate an output of $3.9–11.1 trillion in nine different surroundings, including retail, workshops, and towns, beginning in 2025, and the sum of IoT devices is projected to increase to 754 100 million, which is correspondent to adding 127 devices/sec beginning in

DOI: 10.1201/9781032648309-7

2020 [2]. The following three phases summarize the operation of IoT systems: the positioning of sensors for data collection, the adaptation of collected data into valuable information that can be stowed and retrieved, and the alteration of data to domain knowledge that will be used by the IoT controller for user or system responses. If all tasks comprising the three stages of IoT operations can be automated, an IoT system becomes intelligent [3–5].

Data science is a versatile method of identifying, mining, and giving visions from data through data collecting, data storage and access, data examination, and data communication approaches [6]. Data science is capable of being descriptive, diagnostic, predictive, and prescriptive. It means that administrators can utilize data science to govern what occurred, why it occurred, what transpired, and what they should do given projected consequences. Because the automation of an intelligent IoT system necessitates all data science duties, data science will be the best candidate technology able to address the difficulties that intelligent IoT systems confront [7]. The first major issue for intelligent IoT is how to gather data for IoT application features, as well as how to design sensor placement and connectivity through communication systems or networks. The next stage is to figure out how to use machine learning or artificial intelligence (AI) algorithms [8] to examine and understand the obtained intelligent IoT data. Lastly, it is critical to properly interconnect analytical results with the users of intelligent IoT strategies or devices.

We believe that the succeeding types of challenges should advantage from data science-related skills in the development of upcoming intelligent IoT schemes. The first issue is determining how to cope with the intelligent IoT data. The sum of data generated in each application unit of the intelligent IoT system is at least on the terabyte (TB) scale. Because the processing and communication capabilities of intelligent IoT devices are relatively restricted, collecting, exchanging, storing, and accessing such a massive volume of data at an intelligent IoT device is an extremely difficult operation [9–11]. Deep learning is a game-changing technology in machine learning and AI [12]. The applications of deep learning on IoT devices are frequently required to operate in real time. For instance, security camera-based object-recognition jobs often need a discovery dormancy of less than 400 ms to collect and reply to target events in a quick response time, such as aberrant targets (recognized by deep learning technology) coming within a building.

The existing IoT devices frequently offload intelligent computing to the cloud platform. But, constant and dependable wireless communication associations, which are only accessible in a few areas and at a high charge, represent one of the key challenges for these intelligent IoT devices to meet real-time needs. As a result, the second class of intelligent IoT difficulties is the lack of powerful machine learning and AI algorithms capable of carrying out data analysis with input data influenced by unstable communication lines [13,14]. But, providing machine learning and AI abilities on the intelligent IoT device side is a difficult task. Intelligent IoT devices have modest memory sizes, have minimal power consumption, and are distributed. The third group of difficulties is to create new machine learning and AI algorithms that can be distributedly applied to IoT devices with limited memory and power [15,16]. Finally, for any new technology, the trust, security, and privacy of intelligent IoT operators are constantly top priorities. With such many intelligent IoT devices, it is difficult to apply data science to improve trust management, access control systems,

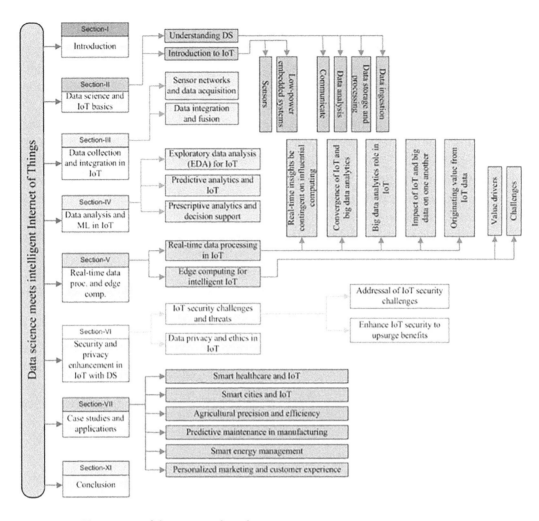

FIGURE 5.1   Taxonomy of the proposed work.

and safe data sharing while keeping privacy in mind across intelligent IoT schemes. The taxonomy of this chapter is illustrated in Figure 5.1.

The rest of this chapter is ordered as follows: Section 5.2 presents the basics of data science and IoT, while Section 5.3 presents data collection and integration in IoT. Section 5.4 describes data analysis and machine learning in IoT, whereas Section 5.5 explains real-time data processing and edge computing. Furthermore, Section 5.6 focuses on security and privacy enhancement in IoT with data science, and Section 5.7 presents case studies and applications. Finally, this chapter concludes in Section 5.8.

5.2 DATA SCIENCE AND IoT BASICS

The most basic definition of data science is basically the study of procedures/events that help in the extraction of value from existing data [17]. In the context of IoT, data denote the information generated by devices, sensors, applications, and smart devices. At the same time, value infers predicting future patterns and outcomes based on such existing data.

5.2.1 Understanding Data Science

Data science is a multidisciplinary field that integrates statistical analysis, machine learning, data visualization, and domain knowledge to derive meaningful insights from data [18,19]. This section digs into the fundamental principles of data science and their importance in generating data-driven decisions.

Data science combines statistics and math, advanced analytics, specialized programming, AI, machine learning, and subject matter knowledge to reveal important visions unseen in an institution's data [20]. These revelations can be utilized to direct decisions and deliberated arrangements. Data science is one of the fastest-growing topics in every organization due to the growing size of data sources and data. Consequently, it is no astonishment that [21] declared the position of data scientist to be the "stimulating occupation of the twenty-first century." Administrations are progressively relying on them to comprehend data and deliver relevant endorsements to advance business upshots. The data science lifespan encompasses numerous tools, roles, and procedures that enable experts to generate illegal visions. A data science project involves the following processes frequently:

5.2.1.1 Data Ingestion

Data collection phase of the lifespan begins with the collection of raw structured and unstructured data from all related sources utilizing several approaches. These approaches comprise physical entry, scraping websites, and real-time flooding data from systems and devices. Customer data, for example, can be coupled with unstructured data such as log files, video, audio, pictures, IoT, social media, and other sources.

5.2.1.2 Data Storage and Processing

Because data can come in a variety of formats and structures, businesses must utilize diverse storage solutions based on the sort of data that desires to be taken. The data supervision squads help to develop data storage and organization standards, which facilitates processes involving machine learning, analytics, and deep learning models [22–24].

5.2.1.3 Data Analysis

Data scientists do investigative data analysis in this case to look for trends, biases, ranges, and value distributions in the data. This data analytics research leads the formulation of hypotheses for a/b testing. It also permits specialists to identify the data's significance for use in analytical analytics, machine learning, and deep learning modeling efforts.

5.2.1.4 Communication

Finally, insights are supplied in the form of information and other data visualizations to assist business experts and other decision-makers in understanding the insights and their impact on the firm. Data scientists can generate visuals using a data science programming language such as R or Python, or they can use specialized visualization tools.

5.2.2 Introduction to IoT

IoT connects a wide range of devices, sensors, and actuators to the internet, allowing them to gather, transmit, and share data on their own [25]. This section gives an overview of the IoT ecosystem, its uses, and the disruptive potential it possesses. IoT is basically the networking of physical goods with electronics incorporated into its architecture that allow them to interconnect and sense connections with each other or with the external world [26,27]. In the next years, the technology of IoT will deliver enhanced degrees of services and will practically change how people live their daily lives. Power, medicine, gene rehabilitations, smart cities, agriculture, and smart hometowns are just a few of the many fields where IoT has a robust occurrence. An overview of IoT and sub-categories is shown in Figure 5.2. IoT is a network of networked computation devices implanted in ordinary objects that can direct and receive data. More than 9 billion "Things" are currently linked to Internet [28]. This quantity is likely to quickly increase to a staggering 20 billion. The primary ones are as follows:

5.2.2.1 Low-Power Embedded Systems
Less battery ingesting and higher presentation are two reverse criteria that influence electrical system design.

5.2.2.2 Sensors Are Very Essential Components in IoT Applications
It is a physical device that detects and measures physical quantities and turns them into signals that another device may process or control.

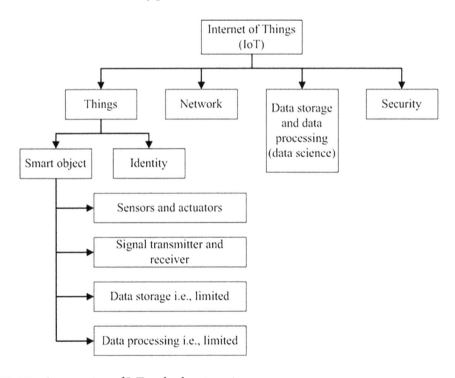

FIGURE 5.2   An overview of IoT and sub-categories.

5.3 DATA COLLECTION AND INTEGRATION IN IoT

Data collection and integration are crucial elements of the IoT ecosystem. IoT devices produce enormous amounts of data from multiple sources; therefore, gathering, filtering, and integrating these data are essential if one is to gain valuable insights and make informed decisions [29]. We will examine the key elements of data collection and integration in IoT in this part.

5.3.1 Sensor Networks and Data Acquisition

IoT devices employ sensors to gather information from the real world. We examine several sensor kinds, their uses, and the problems related to gathering sensor data. In order to capture actual data from the environment, sensor networks are essential parts of the IoT ecosystem [27,30]. These networks are made up of physically connected objects that contain a variety of sensors that gather information on things like temperature, humidity, motion, light, sound, and other things. Data from these sensors are essential for comprehending and keeping track of the physical world, and they also form the basis for data-driven decision-making in IoT applications. IoT devices can independently collect and distribute data across sensor networks, providing valuable information in a range of industries, including manufacturing automation, smart cities, health care, and agriculture. For the IoT to fully realize its promise, revolutionize industries, and enhance our comprehension of the world around us, efficient data collection from sensor networks is essential.

5.3.2 Data Integration and Fusion

IoT devices can produce a wide variety of heterogeneous data. In order to create a comprehensive understanding, the approaches and techniques for integrating and fusing data from multiple sources are covered in this section. Data integration in the IoT requires combining data from various sources, devices, and networks to create a thorough and cohesive view. This connection enables extensive analytics and offers a complete view of the IoT ecosystem. In order to improve accuracy and dependability, data fusion techniques combine data from numerous sensors, creating a more complete picture of the physical world [31].

5.4 DATA ANALYSIS AND MACHINE LEARNING IN IoT

Data analysis and machine learning are essential elements of the IoT environment because they help businesses make intellect of the massive amount of data that IoT devices generate and utilize that data to influence their decisions. Patterns and trends in IoT data can be found using data analysis methods like exploratory data analysis (EDA) and data visualization. Then, using machine learning algorithms, future outcomes are modeled and predicted, enabling predictive maintenance, anomaly detection, and IoT system optimization [32,33]. Combining data analysis and machine learning reveals the right potential of the IoT, turning data into useable intelligence and nurturing invention across commerce, whether it be for resource optimization in smart grids, equipment failure prediction in industrial settings, or personalizing user experiences in IoT applications.

5.4.1 Exploratory Data Analysis (EDA) for IoT

EDA is necessary to comprehend IoT data patterns, trends, and linkages. The significance of EDA and its function in obtaining important insights are highlighted in this section. Realizing the full value of IoT data requires EDA [34]. The IoT ecosystem generates enormous amounts of data, making it crucial to gain insights and understand the underlying patterns and connections. EDA is the process of visually and quantitatively analyzing data to uncover trends, anomalies, and patterns that can be used to drive later data analysis and decision-making processes. EDA provides a comprehensive perspective of the data through data visualization techniques like as scatter plots, histograms, and heatmaps, allowing data scientists and domain experts to spot useful patterns, correlations, and potential issues. EDA improves understanding of IoT data by revealing hidden insights and anomalies, paving the way for data-driven optimizations, predictive modeling, and informed decision-making in a variety of IoT applications and industries.

5.4.2 Predictive Analytics and IoT

Predictive analytics uses previous data to generate educated predictions about future IoT events. We investigate the various machine learning methods utilized in IoT predictive modeling. Predictive analytics and IoT constitute a potent mix that transforms how businesses function and make choices [35]. Predictive analytics in the context of the IoT uses historical data acquired from IoT devices to estimate future trends, behavior, and occurrences. Predictive analytics can find patterns and connections in data by using powerful machine learning algorithms, allowing firms to make accurate forecasts, manage difficulties, and capitalize on opportunities. Predictive analytics, for example, can predict equipment breakdowns in industrial settings, allowing for prompt repair to minimize costly downtime. In healthcare, it can predict patient readmissions, allowing physicians to deliver more individualized care and minimize hospitalization rates [36]. Overall, predictive analytics in IoT optimizes resource allocation, improves efficiency, and enables data-driven strategies across multiple industries.

Predictive analytics and IoT integration not only improve operational efficiency but also provide more proactive and tailored experiences for end users. Predictive analytics may anticipate user preferences, alter settings, and anticipate user demands in consumer IoT applications such as smart homes and wearable devices, resulting in smooth and personalized experiences. Smart home systems driven by IoT, for example, can learn user behavior and automatically alter temperature, lighting, and other preferences. Predictive analytics-enabled wearable devices can analyze health data to deliver timely alerts and personalized health suggestions [37]. Predictive analytics has the potential to alter industries and everyday life as it evolves alongside the expansion of IoT, turning data into actionable insights and making systems smarter, more efficient, and user-centric.

5.4.3 Prescriptive Analytics and Decision Support

Prescriptive analytics goes beyond predictive modeling by offering actionable recommendations and decision support for IoT systems. By offering actionable recommendations and insights, predictive analytics and decision support systems play a critical role in maximizing the value of IoT data. Prescriptive analytics extends predictive analytics by not only forecasting future results but also recommending the optimal course of action to attain desired outcomes. Prescriptive analytics in the context of IoT systems leverages historical and real-time data to optimize decision-making processes. Prescriptive analytics can determine the most effective actions to take in response to changing situations, probable abnormalities, or specified goals by utilizing advanced algorithms and machine learning models. Prescriptive analytics, for example, might prescribe maintenance activities in industrial IoT applications to decrease downtime and increase the lifespan of equipment, thereby enhancing operational efficiency and lowering costs. Prescriptive analytics in healthcare can provide individualized treatment strategies for patients based on their medical history and real-time health data [38]. Prescriptive analytics guarantees that IoT systems function at top performance by providing decision-makers with actionable insights. This results in increased efficiency, productivity, and improved user experiences.

5.5 REAL-TIME DATA PROCESSING AND EDGE COMPUTING

We live in an increasingly digitized and linked world that resembles the future we imagined as children. It is the IoT, a network of physical items that interconnect with one another and data exchange through the Internet using sensors and APIs. Its growth is inevitable, and by 2025, it is anticipated that there will be more than 30 billion IoT networks worldwide, representing an average of nearly four IoT devices per person.

Because of this surge, the amount of data that must be processed and managed is expanding. Traditionally, these networked objects gather data and deliver it to big data centers for processing. However, transmitting the data to the data center for processing takes time that we do not always have, which is an issue in certain use cases where fast reactions are required and every millisecond counts, such as in autonomous driving. To boost agility and efficiency, the edge computing paradigm comes into play [39].

The marine industry, like the remainder of the industrial world, is always growing and looking for new opportunities. Automation, robotics, AI, machine learning, and the digital realm, in general, have changed the game – and the way businesses operate. Smarter ships, automated operations, preventative repair, improved security, and improved supply chain visibility are all prerequisites for meeting the ever-increasing demands for profitability, efficacy, and cost-efficiency, and many of these major developments are direct results of the Industrial IoT (IIoT)'s introduction around the turn of the millennium.

Edge computing technology has several benefits, including reduced latency, higher performance, and increased reliability. It does, however, provide several issues, including security and the administration of remote computing resources. Before implementing edge computing technology, businesses should thoroughly examine their needs. Edge computing technology, with its capacity to provide real-time processing and data analysis, is composed to play a big role in the future of computing.

5.5.1 Real-Time Data Processing in IoT

In IoT situations, real-time insights are frequently required for quick decision-making. We talk about real-time data processing algorithms and their importance in dealing with streaming data from IoT devices [40]. As IoT use grows, organizations across all industries fight to keep up with the huge datasets that are increasing at an exponential rate. IoT devices and sensors, for instance, may collect gigabytes of data in minutes and that is before you factor in data from CRM, financial reports, social media stations, etc.

Simultaneously, AI, machine learning, and big data analytics are advancing at breakneck speed. Organizations can swiftly extract important evidence from these enormous, diverse data sets and adapt to real-time situations by considering AI for IoT data management and analytics. These technologies, when combined, are enabling game-changing advancements. For example, the intrinsic properties of big data are ideal for rapidly training AI and machine learning systems.

Following that, in real-time, intelligent software can be considered to organize operations, predict equipment failures, and identify security risks. In fully autonomous systems, AI takes control while being directed by a network of interconnected IoT devices.

5.5.2 Real-Time Visions Be Contingent on Influential Computing

Today utmost of the IoT solutions are intended to connect disparate devices inside a network as well as integrate and understand data streams from disparate sources. Such platforms handle numerous challenges posed by IoT, such as security, storing, and interoperability, and they can interact with data analytics tools to give substantial business understanding. However, because utmost data analytics resolutions are built on a cloud computing construction known as Platform as a Service (PaaS), real-time data processing is not imaginable.

Rendering to a fresh report, utilizing cloud-based schemes to handle IoT data has severe restrictions, such as security threats, expectancy, and wasted occasions to act on critical, real-time understandings. Although IoT data streams collect events in real time, processing them entails moving them to the cloud for offline processing and investigation, which can then be examined later. You are also operating in a scheme where you are transferring data to a remote place at a rate that may surpass network bandwidth, squandering storage space and computer resources on meaningless insights.

While only 29% of active companies had edge computing in their analytics strategy, 69% decided that arranging edge computing for IoT data processing will help them attain their key corporate goalmouths. But, it is vital to emphasize that edge computing unaccompanied will not provide real-time data analytics. 5G and Wi-Fi6, IoT platforms, i.e., AWS and Kaa, analytics utensils like Kinesis, Kafka, Spark, Cassandra, Storm, event-driven architectures, and Big Table are all convergent to permit real-time data analytics.

5.5.3 IoT and Big Data Analytics Convergence

AI-driven, IoT, and big data analytics convergence opens a wide range of novel openings for businesses to build more modest commercial replicas. Rendering to Forrester's 2020 Estimates, enterprise policy will play a bigger role in fostering digital transformation. Research reveals that attention in big data has decreased recently, but advances in AI and machine learning are reviving it by offering new approaches to data analysis and usage [41–44]. IoT adoption is simultaneously being fueled by cheaper software, hardware, and sensors as well as expanding values and best practices. As a result, there are increasingly more interconnected "Things" gathering nonstop data streams and metrics to assess machine performance, environmental circumstances, and other variables.

5.5.4 Big Data Analytics Role in IoT

Big data and the IoT are two discrete perceptions, yet they are fetching progressively entwined. You have a vast network of sensors in an IoT environment that collect an extraordinary quantity of data from multiple sources and feed it into the larger big data environment [45]. The amount of data that one of these devices may collect is demonstrated by the following example: Wearable technology called the Oura Ring monitors physical activity, temperature, and sleep. Data are gathered by the device 250 times per second. To put this into perspective, Madison Square Garden would be full in less than 7 hours if one cubic foot of water were to be pumped into it every minute. This information may include, among other things, insights about client usage, sentiment analysis, sales analytics, and behavioral patterns. Big data and IoT work together to deliver contextual visions that can be used to enhance goods, services, and procedures, which will lead to higher sales.

Amorphous IoT data, such as pedestrian activity in gardens, weather forms, or patient well-being, can be combined with other data sources using big data analytics tools to create an inclusive picture of the issue. The data may then be transformed by platforms into useful visions that businesses may utilize to enhance their processes. This implies that social media and consumer behavior insights may be combined with ecological data from sensors, security footage, and geolocation data to build a more complete image of your target audience, carrying them to life in ways that advertising analytics cannot.

5.5.5 Impact of IoT and Big Data on One Another

IoT is gradually boosting value generation in both the public and commercial sectors by improving information flow between people, processes, and the network of interconnected objects, rendering to Carrie MacGillivray, IDC Group V. Leader of IoT, 5G, and Mobility. A data lake is a centralized repository where rare data from sensors and other IoT applications is kept after being created. IoT data, structured data from sources like operation records and consumer profiles, and shapeless data from bases like emails, social media, and logs are all included in these data lakes [46,47].

Reports and visualizations based on the insights drawn from all the data sets added to the data lake can subsequently be produced by big data analytical tools. As a result, this offers a thorough picture of how outside features like market alterations, trends, and environmental factors affect what occurs within a corporation.

5.5.6 Originating Value from IoT Data

Big data and IoT are no longer just gimmicks for hypothetical use cases in the future; they are increasingly establishing themselves as essential instruments for maintaining modesty now. By combining IoT data with already-used commercial tools and external data sets to offer context, they enable enterprises to derive value from IoT systems. The info acquired can subsequently be applied to improve services, amenities, and customer procedures. To make the most of their reserves, businesses must make sure that they have the structure in place to conduct real-time dispensation at gage.

5.5.7 Edge Computing for Intelligent IoT

Edge computing brings computation closer to the data source, dropping latency and enhancing efficiency. In a recent paradigm shift for growing cloud infrastructure, edge or fog computing, cloud computing processing nodes are moved to the periphery of the radio access network for cost-effectiveness and low latency. Additionally, by utilizing computational node resources, edge or fog computing enables intelligent IoT devices and applications to improve quality of experience (QoE), energy productivity, quality of service (QoS), system scalability, reduce transport network loads and circulation, and analyze collected data with low latency [48,49]. Still, there are several issues that must be resolved before edge or fog computing architecture can be fully considered for IIoT applications. These issues include cutting-edge and efficient mobility management systems, smart homes, hospitals, cities, smart cars or unmanned vehicles, active optimization architecture for handling computing and storage resources, intelligent network managing services, privacy, or security issues, balancing the load amoeba, and intel Collaboration between edge/fog computing and cloud computing needs to be investigated more in order to deliver scalable services.

Edge computing, as defined, is a computing model that offers actionable intelligence and insights as near to the point of implementation of IoT as possible, maximizing the projected value propositions of such installations. The reason for this reinterpretation is self-evident: The ability of edge computing to generate value is crucial.

5.5.8 Value Drivers

It is critical to understand the reasoning for such a redefinition that emphasizes value proposition. There are numerous drivers for any IoT product or service. Drivers, as seen in Figure 5.1, can be broadly defined as business, technology, device, or data drivers. All drivers want to make rapid, informed, and actionable judgments. Business drivers, also known as functional or operational drivers, are important to any IoT project's success. Businesses seek to reap benefits from IoT-driven intelligent decisions such as increased productivity, more revenue, improved client satisfaction through novel goods and services, and, most significantly, lives saved. Technological drivers supplement commercial drivers.

5.5.9 Challenges

Edge computing is fundamentally an extension of cloud computing, with the caveat that, due to the inherent mobility requirements of IoT, it is more fragmented and scattered than normal cloud computing. The degree of distribution near IoT locations differs significantly from that on the back-end server. While edge computing, by definition, tries to be as close to IoT devices as possible, businesses continue to utilize edge computing in the same way that they do cloud computing for the following important reasons:

a. The security of edge computing solutions and procedures is still growing, and it is evident that edge devices can be simple targets for hackers.

b. Architecture patterns for edge computing are still emerging. Even though the performance and latency requirements for edge computing differ, data generated by edge devices are still saved in cloud storage systems that are centrally managed.

c. In the traditional cloud computing environment, the memory-intensive, CPU-intensive, and low-latency disk resources required for performing complicated machine learning and deep data processing models are more voluntarily available than at the edge [50–52].

d. Other cloud-based technology abilities and services, such as serverless and succeeded container services, are significantly more mature and cooler to acquire.

As a result of these considerations, data insights and intelligence that drive meaningful IoT decisions must rely on centralized high-performance cloud computing. Cloud computing adds network and processor latency, which is unacceptable in time-critical, sense-and-respond IoT solutions.

5.6 SECURITY AND PRIVACY ENHANCEMENT IN IoT WITH DATA SCIENCE

Enhancing IoT security and privacy with data science is becoming increasingly important as the IoT expands in scope and impact [53]. Data science techniques are critical in strengthening the security of IoT devices and networks while also protecting user privacy. Data science, using the huge amounts of data created by IoT devices, enables the detection of potential security breaches and suspicious behaviors using advanced anomaly detection algorithms. Data science can detect suspicious behaviors such as cyberattacks or illegal access to IoT systems by studying data patterns and user behavior. Furthermore, data science allows for the use of encryption and cryptography techniques to secure data transit and storage, ensuring that sensitive information is kept private and confidential.

Furthermore, data science addresses privacy issues in the IoT context. Data scientists may derive important insights from IoT data while protecting individual user privacy by using privacy-preserving data analysis approaches. Data science can secure users' identity while still allowing relevant analysis by anonymizing and aggregating data. Furthermore, data science is critical in establishing and implementing access control methods that allow only authorized individuals to access specific data, lowering the risk of data breaches. The combination of data science and IoT security and privacy controls ensures that IoT systems are resilient to possible threats and user data is secure, increasing trust and confidence in IoT technology. Data science for IoT and sub-categories is illustrated in Figure 5.3.

5.6.1 IoT Security Challenges and Threats

Because of the networked landscape of IoT devices, security challenges are unique. We investigate potential dangers and vulnerabilities in IoT systems, as well as how data science might assist in mitigating them. IoT devices, as previously noted, were not established with security in attention [54]. Consequently, there are various IoT security problems that can result in disastrous results. In contrast to other technology solutions, IoT security is

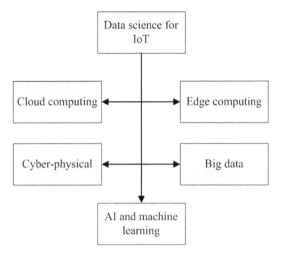

FIGURE 5.3   Data science for IoT and sub-groups.

governed by a few rules and conventions. In addition, most customers are uninformed of the integral dangers of IoT devices [55]. Also, they do not have an idea of how thoughtful the IoT security issues are. Among the many worries about IoT security are listed below:

- **Deficiency of Visibility:** The operators regularly install IoT devices without the information of IT sectors, creating a precise record of what desires to be protected and observed impracticable.

- **Incomplete Security Combination:** Mixing IoT devices into security systems is problematic to unbearable due to their variety and scale.

- **Open-source Code Susceptibilities:** Firmware for IoT devices regularly incorporates open-source software that is disposed to faults and susceptibilities [56].

- **Devastating Data Volume:** Due to the data volume shaped by IoT devices, data management, management, and defense are challenging.

- **Deprived Testing:** Because utmost IoT designers do not list security, they do not undergo behavior-effective susceptibility testing to uncover faults in IoT schemes.

- **Unpatched Susceptibilities:** Numerous IoT devices are unpatched due to various factors, such as a lack of updates and difficulties retrieving and applying patches.

- **Susceptible APIs:** APIs are widely utilized as points of entry into command-and-control (CC) centers, from which SQL injection, man-in-the-middle (MITM), distributed denial of service (DDoS), and network-breaking assaults are launched.

- **Feeble Passwords:** IoT devices commonly come with default passwords that numerous users refuse to update, allowing cyber attackers' easy access. In other cases, users generate passwords that are easily guessable.

5.6.2 Addressal of IoT Security Challenges

A complete technique is necessary to successfully implement and maintain IoT security. It must include a wide range of methodologies and utensils, and consider head-to-head systems, i.e., networks [57]. The following are three important competences for a solid IoT security solution:

- **Learn:** Use safety keys that empower network perceptibility to learn about the environment and the hazard outlines for every set of IoT devices.

- **Protect:** Inspect, monitor, and impose IoT security strategies in conjunction with activity at various locations throughout the structure.

- **Segment:** In the equivalent way that networks are isolated, IoT schemes can be segmented based on strategy clusters and risk outlines.

5.6.3 Enhance IoT Security to Upsurge Benefits

IoT devices are increasingly being used by individuals and organizations. They are not only here to stay, but are multiplying tremendously in new forms. As a result, complexity rises, complicating efforts to manage IoT system security successfully. IoT security concerns range from repelling hostile insiders to protecting against nation-state attacks [58]. Because of the inherent susceptibility of IoT devices and the volume of their deployment, attacks continue to grow and scope. Regardless of IoT security concerns, securing IoT devices is a worthwhile expenditure. To compete with competing technologies, IoT device value can only be increased by improving security. It lowers risks while improving returns.

5.6.4 Data Privacy and Ethics in IoT

As IoT devices capture large volumes of personal data, privacy and ethical data use become critical. Data privacy and ethics in the IoT have arisen as major issues in the digital era. As the IoT spreads, connecting an increasing number of devices and creating massive amounts of data, the need to protect individuals' personal information and follow ethical values becomes critical [59]. While the seamless integration of gadgets and sensors provides unparalleled ease and efficiency, it also opens the door to potential privacy violations and sensitive data exploitation. Striking a balance between technical progress and consumer privacy has become a difficult task. Adherence to strong data protection legislation and the adoption of robust security measures are critical in ensuring that IoT devices respect the privacy rights of their users. Furthermore, addressing ethical considerations such as data consent, transparency, and accountability in the design and deployment of IoT technologies is critical to establishing user trust and ensuring responsible and ethical use of the vast sums of data generated by these interconnected devices. As the IoT evolves, cultivating a culture of data privacy and ethics will be critical to reaching its full potential while protecting individual freedoms and social well-being.

With the spread of IoT devices all over the world, private info is more reachable than ever in the connected world. The technology of IoT and 5G are powerful and have the potential to revolutionize communication, health care, the supply chain, and other parts.

However, the same technology poses issues for enterprises and governments in terms of defending sensitive data. Furthermore, perspectives on data privacy differ around the world. Government bodies frequently issue regulations that outline the requirements for how data privacy should be handled in a certain location. Compliance with data privacy is determined by these government regulations. In the energy systems business in the United States (US), for example, the Federal Energy Regulatory Commission oversees compliance with privacy regulations. Different governing organizations specify what compliance entails in other locations or businesses in order to guarantee data privacy.

5.7 CASE STUDIES AND APPLICATIONS

Data science and Intelligent IoT have a wide range of submissions in a variety of industries. Here are some case studies demonstrating their real-world applications:

5.7.1 Smart Healthcare and IoT

Investigate how data science is altering healthcare with IoT-enabled medical equipment, remote patient nursing, and individualized treatments. Data science and the IoT are transforming the healthcare business by offering remote patient monitoring and individualized therapies. Wearable health gadgets, for example, can continuously gather data on a patient's activity levels, vigorous signs, and sleep patterns. Data science algorithms are used to examine these data in order to find anomalies, anticipate health hazards, and provide prompt medical interventions. This proactive strategy has greatly improved patient outcomes while also lowering healthcare expenses.

5.7.2 Smart Cities and IoT

Learn how data science is advancing smart city development through intelligent infrastructure, efficient resource management, and data-driven urban planning. Cities all over the world are implementing IoT-enabled technologies to boost efficiency, sustainability, and quality of life. Data science is critical in digesting the massive amounts of data provided by IoT sensors like traffic cameras, weather stations, and waste management systems [60]. City managers may optimize traffic flow, regulate energy use, and plan urban expansion more intelligently by evaluating these data, resulting in more livable and sustainable communities.

5.7.3 Agricultural Precision and Efficiency

Precision agriculture employs IoT devices and sensors to monitor weather patterns, soil situations, and crop well-being. These data are then applied to improve reproduction, irrigation, and pest management, resulting in increased crop yields and resource effectiveness. Intelligent IoT technologies assist farmers in making data-driven decisions, resulting in more sustainable and productive farming operations.

5.7.4 Predictive Maintenance in Manufacturing

IoT devices embedded in industrial equipment collect data on their performance and health in real time. These data are analyzed by data science algorithms to predict probable breakdowns and maintenance requirements. Manufacturers can eliminate unplanned downtime, prolong equipment lifespan, and optimize maintenance schedules by employing predictive maintenance solutions, resulting in significant cost savings and increased productivity.

5.7.5 Smart Energy Management

Data science and the IoT are revolutionizing energy management by providing real-time monitoring and control of energy consumption. Smart meters and IoT sensors capture energy usage data from a variety of sources. Data science algorithms analyze these data to detect patterns and trends, assisting consumers and utilities in optimizing energy usage, forecasting peak demand, and implementing demand-response measures, resulting in decreased energy waste and cheaper costs.

5.7.6 Personalized Marketing and Customer Experience

Retailers and e-commerce platforms collect and analyze customer data from numerous touchpoints using data science and IoT technology. Businesses can offer individualized product recommendations, targeted marketing efforts, and enhanced customer experiences by analyzing customer preferences and behaviors, resulting in higher customer satisfaction and increased sales.

These case studies show how data science and Intelligent IoT are transforming industries, boosting innovation and efficiency, and redefining the way we live, work, and interact with technology. As these technologies advance, their impact on society and industry is expected to become even more profound.

5.8 CONCLUSION

The combination of data science with IoT has opened a new world of options and possibilities. This chapter underlines the revolutionary potential of this synergy and the significance of responsible and ethical data use in order to create a smarter and more sustainable future. As technology advances, integrating data science and the IoT is critical for realizing the full promise of intelligent connection and data-driven decision-making.

REFERENCES

[1] Qin X, Gu Y. Data fusion in the Internet of Things. *Procedia Engineering.* 2011;15:3023–6.

[2] Diechmann J, Heineke K, Reinbacher T, Wee D. *The Internet of Things: How to Capture the Value of IoT.* Technical Report, McKinsey & Company, NY, USA, pp. 1–124, 2018.

[3] Khan WU, Imtiaz N, Ullah I. Joint optimization of NOMA-enabled backscatter communications for beyond 5G IoT networks. *Internet Technology Letters.* 2021;4(2):e265.

[4] Gupta D, Juneja S, Nauman A, Hamid Y, Ullah I, Kim T, Tag eldin EM, Ghamry NA. Energy saving implementation in hydraulic press using industrial Internet of Things (IIoT). *Electronics.* 2022;11(23):4061.

[5] Asif M, Khan WU, Afzal HR, Nebhen J, Ullah I, Rehman AU, Kaabar MK. Reduced-complexity LDPC decoding for next-generation IoT networks. *Wireless Communications and Mobile Computing.* 2021;2021:1–10.

[6] Piccialli F, Cuomo S, Bessis N, Yoshimura Y. Data science for the Internet of Things. *IEEE Internet of Things Journal.* 2020;7(5):4342–6.

[7] Stoyanova M, Nikoloudakis Y, Panagiotakis S, Pallis E, Markakis EK. A survey on the Internet of Things (IoT) forensics: Challenges, approaches, and open issues. *IEEE Communications Surveys & Tutorials*. 2020;22(2):1191–221.

[8] Helm JM, Swiergosz AM, Haeberle HS, Karnuta JM, Schaffer JL, Krebs VE, Spitzer AI, Ramkumar PN. Machine learning and artificial intelligence: Definitions, applications, and future directions. *Current Reviews in Musculoskeletal Medicine*. 2020;13:69–76.

[9] Lockl J, Schlatt V, Schweizer A, Urbach N, Harth N. Toward trust in Internet of Things ecosystems: Design principles for blockchain-based IoT applications. *IEEE Transactions on Engineering Management*. 2020;67(4):1256–70.

[10] Tufail AB, Ullah I, Khan R, Ali L, Yousaf A, Rehman AU, Alhakami W, Hamam H, Cheikhrouhou O, Ma YK. Recognition of ziziphus lotus through aerial imaging and deep transfer learning approach. *Mobile Information Systems*. 2021;2021:1–10.

[11] Yousafzai BK, Khan SA, Rahman T, Khan I, Ullah I, Ur Rehman A, Baz M, Hamam H, Cheikhrouhou O. Student-performulator: Student academic performance using hybrid deep neural network. *Sustainability*. 2021;13(17):9775.

[12] Ghosh A, Chakraborty D, Law A. Artificial intelligence in Internet of Things. *CAAI Transactions on Intelligence Technology*. 2018;3(4):208–18.

[13] Mazhar T, Irfan HM, Haq I, Ullah I, Ashraf M, Shloul TA, Ghadi YY, Imran, Elkamchouchi DH. Analysis of challenges and solutions of IoT in smart grids using AI and machine learning techniques: A review. *Electronics*. 2023;12(1):242.

[14] Salau BA, Rawal A, Rawat DB. Recent advances in artificial intelligence for wireless Internet of Things and cyber-physical systems: A comprehensive survey. *IEEE Internet of Things Journal*. 2022;9(15):12916–30.

[15] Baccour E, Mhaisen N, Abdellatif AA, Erbad A, Mohamed A, Hamdi M, Guizani M. Pervasive AI for IoT applications: A survey on resource-efficient distributed artificial intelligence. *IEEE Communications Surveys & Tutorials*, vol. 24, no. 4, pp. 2366–2418, 2022.

[16] Mazhar T, Talpur DB, Shloul TA, Ghadi YY, Haq I, Ullah I, Ouahada K, Hamam H. Analysis of IoT security challenges and its solutions using artificial intelligence. *Brain Sciences*. 2023;13(4):683.

[17] Gokhale P, Bhat O, Bhat S. Introduction to IOT. *International Advanced Research Journal in Science, Engineering and Technology*. 2018;5(1):41–4.

[18] Kaur H, Nori H, Jenkins S, Caruana R, Wallach H, Wortman Vaughan J. Interpreting interpretability: understanding data scientists' use of interpretability tools for machine learning. In *Proceedings of the 2020 CHI Conference on Human Factors in Computing Systems*, Honolulu HI, USA, April 21, 2020, pp. 1–14.

[19] Khalil H, Rahman SU, Ullah I, Khan I, Alghadhban AJ, Al-Adhaileh MH, Ali G, ElAffendi M. A UAV-swarm-communication model using a machine-learning approach for search-and-rescue applications. *Drones*. 2022;6(12):372.

[20] Mazhar T, Malik MA, Haq I, Rozeela I, Ullah I, Khan MA, Adhikari D, Ben Othman MT, Hamam H. The role of ML, AI and 5G technology in smart energy and smart building management. *Electronics*. 2022;11(23):3960.

[21] Davenport TH, Patil DJ. Data scientist. *Harvard Business Review*. 2012;90(5):70–6.

[22] Rayes A, Salam S. *Internet of Things from Hype to Reality*. Springer, Switzerland, 2017.

[23] Ullah A, Jami A, Aziz MW, Naeem F, Ahmad S, Anwar MS, Jing W. Deep facial expression recognition of facial variations using fusion of feature extraction with classification in end to end model. In *2019 4th International Conference on Emerging Trends in Engineering, Sciences and Technology (ICEEST)*, Karachi, Pakistan, December 10, 2019, pp. 1–6. IEEE.

[24] Ahmad S, Anwar MS, Ebrahim M, Khan W, Raza K, Adil SH, Amin A. Deep network for the iterative estimations of students' cognitive skills. *IEEE Access*. 2020;8:103100–13.

[25] Ullah A, Wang J, Anwar MS, Ahmad U, Saeed U, Fei Z. Facial expression recognition of nonlinear facial variations using deep locality de-expression residue learning in the wild. *Electronics*. 2019;8(12):1487.

[26] Khan HU, Sohail M, Ali F, Nazir S, Ghadi YY, Ullah I. Prioritizing the multi-criterial features based on comparative approaches for enhancing security of IoT devices. *Physical Communication*. 2023;59:102084.

[27] Mazhar T, Irfan HM, Khan S, Haq I, Ullah I, Iqbal M, Hamam H. Analysis of cyber security attacks and its solutions for the smart grid using machine learning and blockchain methods. *Future Internet*. 2023;15(2):83.

[28] Shafique K, Khawaja BA, Sabir F, Qazi S, Mustaqim M. Internet of Things (IoT) for next-generation smart systems: A review of current challenges, future trends and prospects for emerging 5G-IoT scenarios. *IEEE Access*. 2020;8:23022–40.

[29] Plaza-Hernández M, Gil-González AB, Rodríguez-González S, Prieto-Tejedor J, Corchado-Rodríguez JM. Integration of IoT technologies in the maritime industry. In *Distributed Computing and Artificial Intelligence, Special Sessions, 17th International Conference*, Aquila, Italy, 2021, pp. 107–115. Springer International Publishing.

[30] Li S, Da Xu L, Wang X. Compressed sensing signal and data acquisition in wireless sensor networks and internet of things. *IEEE Transactions on Industrial Informatics*. 2012;9(4):2177–86.

[31] Ding W, Jing X, Yan Z, Yang LT. A survey on data fusion in internet of things: Towards secure and privacy-preserving fusion. *Information Fusion*. 2019;51:129–44.

[32] Abideen ZU, Mazhar T, Razzaq A, Haq I, Ullah I, Alasmary H, Mohamed HG. Analysis of enrollment criteria in secondary schools using machine learning and data mining approach. *Electronics*. 2023;12(3):694.

[33] Haq I, Mazhar T, Malik MA, Kamal MM, Ullah I, Kim T, Hamdi M, Hamam H. Lung nodules localization and report analysis from computerized tomography (CT) scan using a novel machine learning approach. *Applied Sciences*. 2022;12(24):12614.

[34] Fagroud FZ, Ajallouda L, Toumi H, Achtaich K, El Filali S. IOT search engines: Exploratory data analysis. *Procedia Computer Science*. 2020;175:572–7.

[35] Ullah I, Su X, Zhu J, Zhang X, Choi D, Hou Z. Evaluation of localization by extended Kalman filter, unscented Kalman filter, and particle filter-based techniques. *Wireless Communications and Mobile Computing*. 2020;2020:1–5.

[36] Pal R, Adhikari D, Heyat MB, Ullah I, You Z. Yoga meets intelligent Internet of Things: Recent challenges and future directions. *Bioengineering*. 2023;10(4):459.

[37] Rasheed Z, Ma YK, Ullah I, Al Shloul T, Tufail AB, Ghadi YY, Khan MZ, Mohamed HG. Automated classification of brain tumors from magnetic resonance imaging using deep learning. *Brain Sciences*. 2023;13(4):602.

[38] Ahmad S, Ullah T, Ahmad I, Al-Sharabi A, Ullah K, Khan RA, Rasheed S, Ullah I, Uddin MN, Ali MS. A novel hybrid deep learning model for metastatic cancer detection. *Computational Intelligence and Neuroscience*. 2022;2022, 14 pages.

[39] Ullah I, Qian S, Deng Z, Lee JH. Extended Kalman filter-based localization algorithm by edge computing in wireless sensor networks. *Digital Communications and Networks*. 2021;7(2):187–95.

[40] Simmhan Y, Perera S. Big data analytics platforms for real-time applications in IoT. *Big Data Analytics: Methods and Applications*. 2016:115–35.

[41] Anwar MS, Wang J, Khan W, Ullah A, Ahmad S, Fei Z. Subjective QoE of 360-degree virtual reality videos and machine learning predictions. *IEEE Access*. 2020;8:148084–99.

[42] Ullah A, Wang J, Anwar MS, Ahmad A, Nazir S, Khan HU, Fei Z. Fusion of machine learning and privacy preserving for secure facial expression recognition. *Security and Communication Networks*. 2021;2021:1–2.

[43] Javaid S, Javaid N, Alhussein M, Aurangzeb K, Iqbal S, Anwar MS. Towards efficient human-machine interaction for home energy management with seasonal scheduling using deep fuzzy neural optimizer. *Cognition, Technology & Work*. 2023;25:1–4.

[44] Anwar MS, Wang J, Ahmad S, Khan W, Ullah A, Shah M, Fei Z. Impact of the impairment in 360-degree videos on users VR involvement and machine learning-based QoE predictions. *IEEE Access*. 2020;8:204585–96.

[45] ur Rehman MH, Yaqoob I, Salah K, Imran M, Jayaraman PP, Perera C. The role of big data analytics in industrial Internet of Things. *Future Generation Computer Systems*. 2019;99:247–59.

[46] Waleed S, Ullah I, Khan WU, Rehman AU, Rahman T, Li S. Resource allocation of 5G network by exploiting particle swarm optimization. *Iran Journal of Computer Science*. 2021;4(3):211–9.

[47] Ge M, Bangui H, Buhnova B. Big data for Internet of Things: A survey. *Future Generation Computer Systems*. 2018;87:601–14.

[48] Ren J, Zhang D, He S, Zhang Y, Li T. A survey on end-edge-cloud orchestrated network computing paradigms: Transparent computing, mobile edge computing, fog computing, and cloudlet. *ACM Computing Surveys (CSUR)*. 2019;52(6):1–36.

[49] Ullah I, Shen Y, Su X, Esposito C, Choi C. A localization based on unscented Kalman filter and particle filter localization algorithms. *IEEE Access*. 2019;8:2233–46.

[50] Ullah A, Wang J, Anwar MS, Whangbo TK, Zhu Y. Empirical investigation of multimodal sensors in novel deep facial expression recognition in-the-wild. *Journal of Sensors*. 2021;2021:1–3.

[51] Ahmad S, Anwar MS, Khan MA, Shahzad M, Ebrahim M, Memon I. Deep frustration severity network for the prediction of declined students' cognitive skills. In *2021 4th International Conference on Computing & Information Sciences (ICCIS)*, Karachi, Pakistan, November 29, 2021, pp. 1–6. IEEE.

[52] Ullah N, Khan MS, Khan JA, Choi A, Anwar MS. A robust end-to-end deep learning-based approach for effective and reliable BTD using MR images. *Sensors*. 2022;22(19):7575.

[53] Aqeel-ur-Rehman SU, Khan IU, Moiz M, Hasan S. Security and privacy issues in IoT. *International Journal of Communication Networks and Information Security (IJCNIS)*. 2016;8(3):147–57.

[54] Hassija V, Chamola V, Saxena V, Jain D, Goyal P, Sikdar B. A survey on IoT security: Application areas, security threats, and solution architectures. *IEEE Access*. 2019;7:82721–43.

[55] Zhou W, Jia Y, Peng A, Zhang Y, Liu P. The effect of iot new features on security and privacy: New threats, existing solutions, and challenges yet to be solved. *IEEE Internet of Things Journal*. 2018;6(2):1606–16.

[56] Khan HU, Ali F, Ghadi YY, Nazir S, Ullah I, Mohamed HG. Human-computer interaction and participation in software crowdsourcing. *Electronics*. 2023;12(4):934.

[57] Karale A. The challenges of IoT addressing security, ethics, privacy, and laws. *Internet of Things*. 2021;15:100420.

[58] Elijah O, Rahman TA, Orikumhi I, Leow CY, Hindia MN. An overview of Internet of Things (IoT) and data analytics in agriculture: Benefits and challenges. *IEEE Internet of Things Journal*. 2018;5(5):3758–73.

[59] AboBakr A, Azer MA. IoT ethics challenges and legal issues. In *2017 12th International Conference on Computer Engineering and Systems (ICCES)*, Cairo, Egypt, December 19, 2017, pp. 233–237. IEEE.

[60] Raza M, Barket AR, Rehman AU, Rehman A, Ullah I. Mobile crowdsensing based architecture for intelligent traffic prediction and quickest path selection. In *2020 International Conference on UK-China Emerging Technologies (UCET)*, Glasgow, UK, August 20, 2020, pp. 1–4. IEEE.

CHAPTER 6

Data Science and Big Data Analytics

Faisal Rehman

University of Mianwali

National University of Science and Technology (NUST)

Muhammad Muneer

University of Mianwali

Muhammad Hamza Sajjad

University of Mianwali

Naveed Riaz

National University of Science and Technology (NUST)

6.1 INTRODUCTION

There is an enormous amount of data processing taking place online. The National Security Agency claims to process an astounding 1,826 petabytes (PB) of data each day [1]. Data generation per day surpassed 2.5 quintillion bytes in 2018 [2]. The amount of created data is expected to double every two years, according to an earlier prediction by the International Data Corporation (IDC) [3]. It has been observed that 90% of the data created worldwide occurred in the previous two years, indicating an even quicker growth rate. Popular platforms such as Google process over 99,000 searches every single second. This makes more than 8.5 billion searches per day. Big data and data science are rapidly expanding fields due to increasing interest and demand. "Data science" has become increasingly popular as a result of its applicability in a variety of sectors, while "big data" has gained traction as businesses realize the benefit of processing and analyzing enormous amounts of data to gather knowledge and make wise decisions. Figure 6.1 shows a comparative graph of data science and big data trends.

DOI: 10.1201/9781032648309-8

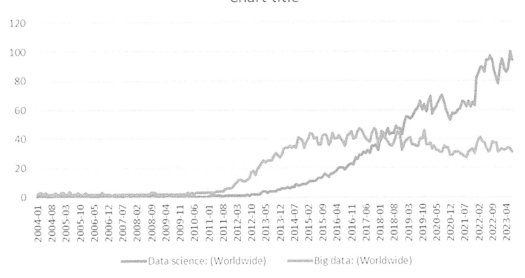

FIGURE 6.1   A comparative graph of data science and big data trends.

To extract meaningful insights from enormous datasets, advanced data analysis techniques are essential for transforming big data into smart data [4]. Smart data provides organizations and businesses with useful information that improves their ability to make decisions. Healthcare professionals can be empowered to provide patients with efficient and cost-effective solutions by evaluating large datasets from programs like Clinical Decision Systems and Electronic Health Records [5]. Better judgments may be made by looking at overall patient history patterns as opposed to merely relying on localized or recent data. The 17 Vs of big data, which are validity, velocity, versatility, volume, veracity, value, variety, voluntariness, volatility, venue, vagueness, viscosity, variability, visualization, vocabulary, virality, and verbosity [6], provide difficulties for traditional data analytics when dealing with big data analysis. These characteristics might cause traditional data analytics to lose their efficacy. Advanced data analysis methods are needed to extract valuable insights from huge and complicated datasets to address these difficulties. Several artificial intelligence approaches, including deep learning, natural language processing (NLP), data mining, machine learning (ML), and expert systems, have been developed to solve the difficulties of large data analytics [7]. For managing enormous amounts of data, these approaches provide quicker, more accurate, and more exact solutions [8]. These cutting-edge analytical methods' main goal is to unearth important data, obscure patterns, and undiscover relationships inside huge datasets. For instance, by carefully examining past patient data, it is possible to spot potentially harmful illnesses early, which might result in better treatment strategies or a cure [9]. Additionally, complex business decisions like entering new markets or launching new goods can be supported by simulations with increased decision-making capabilities, resulting in better decisions. The broad discipline of deriving information and insights from data is covered by data science. Big Data refers

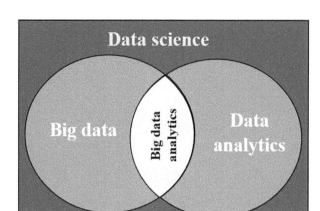

FIGURE 6.2 The relationship between data science, big data, big data analytics, and data analytics.

to huge and complicated datasets, whereas Data Analytics and Big Data Analytics both include the study of data, with the latter addressing data analysis more generally and the former concentrating primarily on large-scale datasets shown in Figure 6.2.

AI-powered big data analytics holds tremendous potential; however, it also brings multiple drawbacks, especially when confronted with uncertainty. There are seven important big data Vs—velocity, verbosity, variety, versatility, volume, veracity, and value, all introduce different types of uncertainty that can affect the quality and reliability of data. Some types of uncertainty in big data include measurement uncertainty, incompleteness, and imperfect, or noisy data. The whole analytics process, including data gathering, organization, and analysis, can be affected by uncertainty. The majority of data mining and ML approaches face a significant hurdle when handling ambiguous and incomplete data. Furthermore, biased training data may prevent ML algorithms from producing optimal results. According to Wang et al. [10], uncertainty considerably influences how well big data analytics function, highlighting six primary problems. Dealing with the uncertainty present in large datasets is a different issue, and data mining and ML algorithms frequently face similar difficulties.

These uncertainties might exacerbate any mistakes or shortfalls in the whole analytics process when grown up to the big data level. Because uncertainty may greatly affect the accuracy of an automated technique's output, minimizing uncertainty in big data analytics becomes a primary concern for all automated techniques. Uncertainty management must be done well for big data analytics to produce trustworthy and valuable insights.

Our review of the literature revealed that limited study to understand how uncertainty influences the fusion of big data and applied analytics methodologies. This chapter provides a summary of current AI methods in the context of big data analytics, including NLP, Computational Intelligence (CI), and ML, to fill this gap [11]. A particular emphasis is placed on the issues posed by uncertainty. The report also suggests future fields of inquiry for this type of study.

This chapter has the following contributions:

- We investigate how different variables affect the crucial Vs of big data features in data science and big data analytics.

- We examine how each of the different big data analytics methods is impacted by uncertainty.

- We go over many approaches that may be used to successfully address the problems that uncertainty presents.

6.2 BIG DATA

Big data was recognized as a critical component for productivity, innovation, and competition in May 2011 [12]. In 2018, over 3.7 billion people used the Internet, representing an increase of 7.5% from 2016 [2]. At its peak in 2010, more than 1 zettabyte (ZB) of data was produced, and this number increased to 7 ZB by 2014 [13]. Velocity, volume, and variety were the three criteria used to describe big data at its inception in 2001 [14]. In 2011, IDC expanded the definition of big data by adding a fourth criterion, value [15]. Veracity was introduced as the fifth big data attribute in 2012 [16]. Although several more traits are identified by various Vs, this chapter focuses on the seven most prevalent traits, as indicated in Figure 6.3.

6.2.1 Big Data Characteristics

6.2.1.1 Volume

Volume refers to the extensive amount of data that is continuously generated, relating to the magnitude and extent of the information collected. It is difficult to establish a consistent standard for what constitutes a "big dataset" since it relies on several factors, including

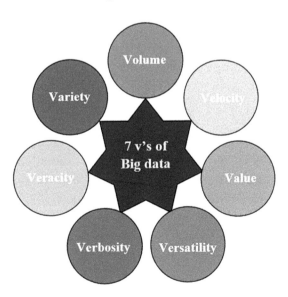

FIGURE 6.3　The seven most common properties of big data.

the type of data being analyzed and the time period. Big data is now typically categorized as datasets in the Exabyte (EB) or Zettabyte (ZB) categories [8]. However, smaller datasets still provide difficulties. For instance, Walmart gathers 2.5 petabytes (PB) of data per hour [17] from more than a million consumers, which might lead to scalability and uncertainty problems. Managing exceedingly large datasets can be a challenge for database tools. Moreover, many current data analysis methodologies are not tailored for handling massive databases and may prove inadequate when processing and comprehending information at such vast scales [8,18].

6.2.1.2 Variety

The term "variety" is used to describe the many types of data that can be encountered within a big data dataset, including unstructured, semi-structured, and structured information. Structured data, similar to the information stored in relational databases, are well organized and easy to categorize. In contrast, unstructured data, such as text and multimedia content, lack a predefined format and are difficult to evaluate. Semi-structured data, often found in NoSQL databases, use tags to distinguish between different data elements but rely on database users to enforce the structure [19]. Uncertainty arises when dealing with mixed data types, converting between multiple data formats, or when the underlying dataset structure changes over time. Traditional big data analytics algorithms face difficulties when handling multi-modal, imperfect, and noisy data from various sources.

It might be difficult to effectively analyze unstructured and semi-structured data since they originate from many sources and have various forms and representations. In real-world databases, the analysis process is significantly impacted by inconsistent, short, and noisy data. Data cleaning, data integration, and data transformation are only a few of the data preparation approaches used to solve the problems brought on by the diversity in large data, such as noise and inconsistent data. The goals of data cleaning procedures are to enhance data quality and reduce data elements. These methods can greatly improve data analysis performance by eliminating noisy components throughout the analysis process. For instance, data cleaning can enhance classification accuracy in ML by identifying and removing mislabeled training examples, which can assist in discovering and repairing problems [20].

6.2.1.3 Velocity

The term "velocity" in the context of big data refers to the speed at which data are processed. This velocity can be categorized into real-time, streaming, and near-real-time. It is crucial that the processing speed matches the rate of data generation [8]. Devices connected to the Internet of Things (IoT) produce sensor data continually [21]. Delays in processing and delivering these data to relevant parties, especially when it contains critical medical information, can have severe consequences, such as patient harm or even loss of life, as exemplified by a pacemaker reporting a medical emergency [16]. Similarly, cyber-physical devices rely on real-time operating systems that adhere to exact timing specifications. As a result, when data from a big data application are not supplied on time, issues may arise.

6.2.1.4 Veracity

Veracity relates to the data's accuracy, which might be unclear or uncertain. For example, according to IBM's estimate, poor data quality costs the US economy approximately $3.1 trillion a year [22]. The quality of data is often assessed as good, poor, or undefined due to inconsistencies, noise, ambiguity, or incompleteness. Building accuracy and confidence in big data analytics gets increasingly difficult as data sources and types become more diverse. For example, data from platforms like Twitter may pose issues because users sometimes mix their personal opinions with official company information on the same account. Using methods created for Twitter datasets might result in problems because of this. Similarly, ambiguity or irregularities in the dataset might impair the accuracy of the analytics process when evaluating millions of records in the healthcare industry to identify disease patterns for breakout mitigation [22].

6.2.1.5 Value

In contrast to the preceding Vs, which mostly concentrated on the difficulties presented by big data, "value" refers to the relevance and utility of data for making informed decisions. Notable businesses that have successfully tapped into the potential of big data through analytics in their products include Facebook, Google, and Amazon. To provide users with individualized product suggestions and increase sales and user engagement, Amazon, for instance, analyzes large user datasets and purchase data. Google improves location services in Google Maps by using location information provided by Android users. In a similar vein, Facebook uses user activity to deliver relevant ads and friend suggestions. These businesses have shown great success by deriving insightful knowledge from sizable raw data collections, which has helped them make wiser business decisions [23].

6.2.1.6 Verbosity

Big data is a massive collection of information from several sources that can be well organized or poorly structured and contain both accurate and inaccurate information. Bad data is inaccurate or lacking, and storing it might be dangerous. It is crucial to ensure that the data we retain are secure, relevant, comprehensive, and reliable to avoid wasting storage space and processing time. Early in the data management process, it's crucial to employ appropriate procedures to determine the worthiness of the information, thus conserving valuable resources. Big data has a trait known as "verbosity," which refers to the possibility of repeating or overlapping information from numerous sources, making it difficult to efficiently manage and analyze.

6.2.1.7 Versatility

Big data is becoming increasingly vital for several enterprises, academics, and governments. It is used for a variety of things, including urban planning, environment modeling, visualization, analysis, environmental security, quality categorization, computational analysis, comprehending biology, and developing and manufacturing processes. It also aids in the development of affordable models and the tasteful exploration of outcomes. Big data has this ability to be resourceful and adaptable, which is why we call it "versatility," which means that it may be used flexibly for many contexts and objectives.

6.3 UNCERTAINTY IN BIG DATA

Uncertainty, in general, describes a circumstance where information is unreliable or uncertain [24]. Every stage of big data learning involves some level of uncertainty [25], and these uncertainties can come from a variety of sources, such as data collection (such as environmental variations and sampling problems), concept variance (where analytics objectives may differ), and multimodality (involving complex and noisy patient health records with numerical, textual, and image data). Due to noise and incompleteness, for instance, many attribute values relating to the time of large data events may be missing. In social networks, between 80% and 90% of connections between data points are often absent, and over 90% of attribute values in patient reports, which are transcriptions of physician diagnoses, may be missing [26]. According to an IBM study from 2014, by 2015, 80% of the data in the globe, according to industry analysts, will be suspect.

The accuracy and efficacy of the results might be negatively impacted by a variety of uncertainties in the field of big data and big data analytics. When training data are utilized in learning algorithms, it may produce incorrect results if it is incomplete, biased, or the result of poor sampling. To overcome these challenges, it's crucial to enhance big data analytical methods to effectively manage uncertainty. There have been a lot of meta-analysis studies recently that incorporate uncertainty and data-driven learning. The effectiveness of learning from big data is substantially impacted by how uncertainty is managed throughout the whole data analytics process. According to an additional study, big data has two distinct properties that are not present in small-sized data. These properties include multimodality (which involves multiple data kinds) and altered uncertainty (which involves uncertainty modeling and measurement for huge data). The size of the dataset is strongly connected with data processing and data uncertainty. To combat ambiguous or incorrect information, techniques like fuzzy groups may be used to describe uncertainty in huge datasets [27]. Big data may also contain hidden linkages, which raises the level of uncertainty and complicates analysis and interpretation.

As a result, assessing uncertainty in huge datasets can be difficult, especially when it's possible that the data were biased when they were acquired. There are many different sorts of uncertainty, and several theories and methods have been created to model them. We next go over a few typical methods.

The Bayesian theory subjectively interprets probability to explain a rational agent's confidence levels in ambiguous propositions, which depend on prior knowledge and practical experience. The belief function theory offers a framework for combining fragmented data in uncertain settings, while probability theory deals with the statistical properties of input data, including unpredictability. The classification entropy evaluates the degree of uncertainty between classes and provides a confidence index. While values closest to one suggest participation in numerous classes, values closer to zero show definite categorization in a single class. Fuzziness assesses uncertainty in categorical terms, particularly those used in the human language such as good and bad. To deal with uncertainty in the real world, fuzzy

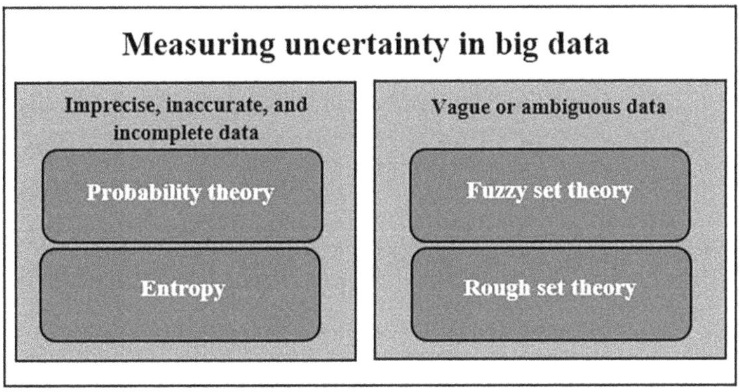

FIGURE 6.4   Big data uncertainty measurement.

logic employs an approximation reasoning technique that approximates human thinking. Fuzzy logic handles uncertainty connected to human perception. Shannon's entropy measures the average quantity of data lacking in a random source to quantify the quantity of information in a variable. Instead of using precise notions to describe concepts, the rough set theory uses two approximations (upper and lower) to deal with ambiguous, unclear, or missing information. Rough theory and the fuzzy set deal with hazy or confused data, whereas Shannon's entropy model and the theory of probability deal with imperfect, false, and inaccurate data. For an illustration of these techniques, see Figure 6.4.

Big data analytics is one of the most important steps in determining the degree of uncertainty. The accuracy of the results might be significantly impacted by neglecting uncertainty in the data or the analysis approach, even though there are many ways to examine large data. Big data analytical approaches may be used with uncertainty models like probability theory, fuzziness, and rough set theory to increase the precision and significance of the outcomes. Common techniques for modeling uncertainty and decision-making in previous research include fuzzy set theory and Bayesian models. Table 6.1 contrasts and condenses the relevant approaches, with a focus on probability theory, rough set theory, fuzzy set theory, and Shannon's entropy. It also compares various uncertainty management techniques. Table 6.1 contrasts and summarizes the methods that we have determined to be pertinent, with a particular emphasis on probabilistic theory, rough set theory, Shannon's entropy, and fuzzy set theory.

6.4 BIG DATA ANALYTICS

The act of big data analytics is looking over enormous databases to seek patterns, previously unrecognized links, consumer preferences, market trends, and other essential data that were difficult to analyze using traditional technologies. Analysis methodologies have to be re-evaluated to get around their processing time and space constraints when the 17 Vs characteristics of big data have been established. In today's digital data world, the applications for big data are continually growing. Big data technologies and services are expected to rise by around 36% annually on a worldwide scale between 2014 and 2019, whereas revenue from business analytics and big data is predicted to grow by more than 60% [32].

TABLE 6.1 Uncertainty Management Techniques

Uncertainty Models	Features
Probability theory	Powerful for dealing with subjective uncertainty and unpredictability where accuracy is needed
Bayesian theory Shannon's entropy	Ability to handle complicated data [28]
Fuzziness	Handles incorrect and confusing info in hard-to-model schemes Accuracy is not assured Easy to learn and utilize [28]
Function of belief	Situations have to be addressed with some innocence Calculates the probability of a certain hypothesis by combining several sorts of evidence Takes into account all supporting data for the hypothesis Ideal for data that are complicated and partial Mathematically challenging but helps reduce ambiguity [28]
Rough set theory	Offers a method of investigation that is objective [29] Addresses data ambiguity Minimum information required to identify set membership Utilize just the details offered by the provided data [30]
Classification entropy	Eliminates vagueness between the classifications [31]

Indeed, several cutting-edge data analysis methods, including data mining, NLP, ML, and CI, are crucial in the field of big data analytics. To break down complicated big data issues into smaller, more manageable jobs, other possible tactics including instance selection, parallelization, sampling, incremental learning, feature selection, granular computing, and divide-and-conquer are used. Firms may maximize the advantages of big data by implementing these tactics and strategies, which help firms make more informed choices, save operating costs, and allow more effective data processing.

- Parallelization is a potent tool used in big data analytics that shortens computing times by splitting up complex issues into simpler ones and carrying out these smaller jobs concurrently. This entails splitting up the job among several threads, cores, or processors, so they may focus on various aspects of the data concurrently. Because smaller activities are executed concurrently rather than sequentially, parallelization considerably accelerates the data processing process, increasing the overall effectiveness of big data analytics [10]. Parallelization is an essential strategy for managing large-scale data processing because, while it doesn't lower the quantity of work performed, it does effectively shorten the time required to finish the analysis.

- The divide-and-conquer tactic is a vital method for handling huge data. It entails three steps: (1) breaking a huge issue down into minor issues; (2) solving each of the minor issues independently in a way that helps to solve the larger overall problem; and (3) combining the smaller problem solutions to come up with a solution for the entire large problem. For many years, this method has been used extensively in managing enormous databases because it enables the editing of records in groups rather than the processing of all the data at once. Divide-and-conquer techniques make big data processing easier to control and more effective, allowing for faster analysis of massive datasets with less computing work.

- A popular learning approach for streaming data is incremental learning. Traditional batch learning trains the model using historical data, whereas incremental learning only uses fresh data. This indicates that the algorithm changes its parameters throughout time, integrating fresh input data just once for training purposes [10].

- By selecting, modifying, and investigating a smaller portion of the data, sampling is a data reduction technique used in big data analytics to detect trends in massive datasets [10]. The factors employed for data selection affect the efficacy of sampling.

- A huge space's constituents are broken down into smaller sets, or granules, using granular computing. By condensing several huge objects into a more manageable and compact search area, this method efficiently characterizes uncertainty in the search space.

- A common strategy for dealing with huge data is feature selection, which aims to pick a subset of pertinent characteristics for a more accurate data representation. It turns out to be an effective data mining technique for creating large datasets.

- In many ML and data mining jobs that include data preprocessing, instance selection is a useful feature. Instance selection can be used to reduce the number of training sets and the length of the categorization or phase of training.

To create resilient and high-performing systems, it is essential to consider the expenses of uncertainty (both monetary and computational) and the difficulties in creating efficient models for uncertainties in big data analytics. In the part that follows, we will go through several unresolved concerns related to how uncertainty affects big data analytics.

6.5 AL APPROACHES IN BIG DATA ANALYTICS AND UNCERTAINTY

In this section, three AI techniques are often used in big data: (1) NLP, (2) ML, and (3) CI, and the effects of uncertainty are investigated. While there are several alternative analytics methods, our primary focus is on these three techniques. We examine the underlying uncertainties that each methodology encounters and discuss the techniques and approaches to reduce this uncertainty.

6.5.1 Big Data and Machine Learning

To improve data-driven decision-making, ML is often used in data analytics to develop models for knowledge discovery and prediction. Traditional ML techniques, on the contrary, are not computationally effective or scalable enough to manage big data's unique properties, including its volume, speed, variety of kinds, incompleteness, low-value density, ambiguities like biased practice data, and surprise data types. For large data analysis, several advanced ML approaches, such as deep learning, feature learning, transfer learning, active learning, and distributed learning, have been suggested to meet these issues.

Techniques used in feature learning allow a system to robotically find the illustrations required for identifying and categorizing features from raw data. The choice of data format has a big impact on how well ML algorithms perform. Deep learning algorithms were created expressly to sift through and extract useful information from the large volumes of data gathered from multiple sources, including the minute differences found within a picture, such as differing materials, lighting conditions, and forms. However, because of their complexity and the vast amounts of data they handle, current deep learning models have the issue of incurring a significant computational cost. Distributed learning is a technique that may be used to scale up the learning process and overcome the scalability problem with classical ML. By properly transferring knowledge from a related area, transfer learning effectively improves a learner's performance by using knowledge obtained in one domain and applying it to new situations. The use of adaptive data collection approaches by active learning, on the other hand, addresses labeling issues and accelerates ML processes by automatically modifying settings to acquire the most valuable data quickly. The main sources of the uncertainty problems that ML approaches encounter include learning from data that have poor accuracy (i.e., incomplete and uncertain data) and small value (i.e., irrelevant to the present situation). Other ML methods, such as fuzzy logic theory, deep learning, and active learning, have shown special usefulness in addressing the issue of reducing uncertainty [33]. The efficiency of various methods for dealing with uncertainty in ML problems is shown in Figure 6.5. Machine learning (ML) can be affected by

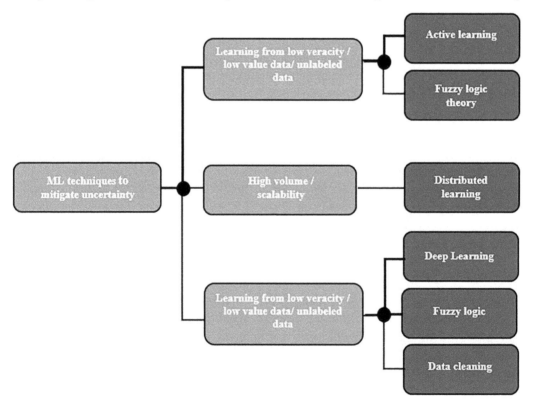

FIGURE 6.5 Machine learning methods address uncertainty in massive data.

uncertainty in several ways, including when dealing with inadequate or inaccurate training samples that result in ambiguous categorization boundaries and a restricted comprehension of the target data. Data may occasionally lack labels, which present a problem for conventional ML methods. Learning from unlabeled data is challenging because imprecise instructions can produce confusing outcomes, and manually labeling huge datasets can be costly and time-consuming. By picking the most important examples for labeling, active learning has proven successful in resolving this problem and improving the effectiveness and efficiency of the learning process. Another ML approach that can address concerns with consistency and incompleteness in the classification process is deep learning [18]. It allows for the automated extraction of pertinent characteristics from data, minimizing the requirement for manually created features and enhancing the model's resilience. Another method for modeling uncertainty effectively is fuzzy logic theory. By adding a fuzzy relationship to each SVM input idea, fuzzy support vector machines (FSVMs) allow for a more flexible and adaptive classification process. Fuzzy logic helps FSVMs manage uncertain and imprecise input effectively, improving performance in classification tasks. Due to the flexibility offered by fuzzy logic's application in the learning process, support vector machines (SVMs) are improved by lessening the effect of data point noise. In light of this, even while insecurity is a substantial badly behaved for ML algorithms, including efficient methods to quantify and describe insecurity can lead to organizations that are more adaptable and successful. ML models may produce more accurate and trustworthy results by addressing uncertainty effectively, improving their overall performance across a range of applications.

6.5.2 Big Data and Natural Language Processing

Devices can analyze, understand, and even produce text through NLP, which is based on ML [8]. In the field of big data analytics, NLP works with enormous volumes of text data and may quickly uncover insightful information. Lexical acquisition, word sense disambiguation, and part-of-speech (POS) tagging are examples of popular NLP techniques. Using NLP-based methods, several text-mining tasks have been successful, including info extraction, text summarization, classification, question answering, clustering, opinion mining, and topic modeling [8]. For instance, in economic and scam investigations, NLP techniques like named entity abstraction and info retrieval may effectively sift through enormous volumes of textual material, assisting in the discovery of criminal identities and pertinent bank records to help fraud investigations. By identifying semantic similarities, NLP can assist in building and recovering traceability relationships between textual artifacts. Big data and NLP may also be used in conjunction to anticipate changes in the composite stock price index and evaluate news chapters. Organizations may use NLP in big data analytics to extract insightful information from massive textual data sources, facilitating improved decision-making and increasing a variety of applications in practical settings.

Big data analytics presents NLP with a variety of issues due to uncertainty. One illustration is the effect of keyword search, a text-mining technique frequently used to control enormous volumes of written data. A reserve word search includes entering a list of pertinent words or idioms and looking for instances of these words (i.e., search keywords) in the intended dataset (e.g., a database or document). The accuracy of keyword searches might be impacted by uncertainty, though, as the inclusion of a term does not ensure the relevancy of the content. For instance, a keyword search may miss words with spelling mistakes that are nevertheless significant since they frequently match precise strings. To overcome this, fuzzy search technologies and Boolean operators provide greater flexibility by letting the search for words that have the intended spelling but are spelled differently. Although a useful strategy in text mining, keyword or keyphrase search has its drawbacks. A limited choice of search keywords might result in the omission of crucial information, while a larger set could provide a significant number of unrelated false positive results. Automatic POS taggers provide yet another noteworthy illustration of how uncertainty impacts NLP. In some terms, such as "bimonthly," which might indicate depending on the situation, every two months or twice a month, these taggers run into ambiguity. Furthermore, terms like "quite" could have distinct connotations for audiences in the United States and the United Kingdom. Additionally, there may be difficulties with correct tagging and interpretation due to the ambiguity of periods ('.'), which can be read as either punctuation (such as a full stop) or a token (such as an abbreviation). Utilizing IBM Content Analytics (ICA), according to a recent study, may assist in mitigating these problems. The issue of handling massive amounts of data, meanwhile, has yet to be overcome in this field. Additionally, when working with biological language, which differs much from Standard English, uncertainty and ambiguity have a major influence on part-of-speech (POS) labeling. When employing trained taggers from the Treebank corpus on biological data [34], there are concerns with uncertainty and poor tagging accuracy, according to studies. Stream processing systems are being utilized to handle massive volumes of data fast and with short reaction times to solve this issue. Large-scale textual data may be handled in real time by combining NLP approaches with uncertainty modelings. To properly explore and develop these skills, more study in this field is necessary.

6.5.3 Big Data and Computational Intelligence

Big data analysis relies heavily on CI, which was inspired by natural phenomena [35]. CI methods have proven effective in addressing complex data processing and analytics challenges, especially when traditional techniques fall short due to high complexity and uncertainty. Some commonly employed CI techniques include Artificial Neural Networks (ANN), fuzzy logic, and evolutionary algorithms (EAs). These techniques have been used to solve a variety of search-based issues, including parameter optimization and robot controller optimization. Due to their innate capacity to manage significant quantities of uncertainty, CI approaches are well suited to meet the real-world difficulties of large data. For instance, CI techniques may be used to create models for forecasting user emotions, a challenge that is rife with various unknowns. These models utilize huge informational libraries about human emotion and its intrinsic fuzziness. Despite CI approaches' present capabilities, several issues still exist, particularly when addressing the value and veracity aspects of huge data. New CI strategies that can effectively manage enormous volumes of data and react swiftly to changes in the dataset are therefore being sought after. According to studies [36], swarm intelligence, AI, and ML algorithms can be used to optimize massive data analysis. These methods are used to develop empirical statistical prediction models, collaborative filtering, and train computers to execute predictive analysis tasks. Using CI-based big data analytics solutions makes it feasible to analyze enormous amounts of data more accurately and with less complexity and ambiguity.

By enabling approximation reasoning and modeling of qualitative data using linguistic quantifiers (fuzzy sets), fuzzy logic is an effective strategy for addressing uncertainty concerns in big data analytics. It enables the encoding of ambiguous user-defined and real-world ideas, as well as interpretable fuzzy rules that support inference and decision-making. Data noise presents problems for big data analytics because it causes outlier artifacts and high levels of uncertainty. Fuzzy logic methods effectively manage the data's innate uncertainties. Another study used Map Reduce and fuzzy logic-based matching algorithms for experimental decision provision to do big data analytics, demonstrating significant flexibility in managing data from multiple bases. An additional helpful CI approach for overcoming the difficulties of large data analytics is evolutionary algorithms. By gradually building up a population of potential solutions, EAs imitate the evolution process to find the best answers to challenging issues. Big data is notoriously difficult to analyze because of its huge volume, diversity, and poor degree of authenticity. For instance, utilizing parallel genetic algorithms to analyze medical images has produced successful outcomes in a Hadoop-based system. However, it is crucial to recognize that motion, noise, and unexpected situations may influence the results of CI-based algorithms. Additionally, an algorithm that successfully solves one of these issues may struggle to do so when several variables are in play at once [37].

6.6 OVERVIEW OF THE MITIGATING TECHNIQUES

The study has undertaken a thorough analysis of the different big data analytics methodologies and how susceptible they are to uncertainty. Table 6.2 presents a summary of the

TABLE 6.2   Techniques for Reducing Uncertainty

Artificial Intelligence	Uncertainty	Mitigation
Machine learning	Unfinished training samples An inconsistent classification Using noisy, skewed, and low-quality data to learn Unlabeled data learning Scalability	Fuzzy groups, function selection [38], DL [18], fuzzy sets, and active learning Learning actively Learning that is dispersed [9] Learning in-depth
Natural language processing	Words in the POS An ambiguous classification (simplifying language assumption) while searching for keywords	Fuzzy, Bayesian ICA, LIBLINEAR, and MNB algorithm ICA, open issue [39]
Computational intelligence	Classification (simplifying a linguistic premise) Complex, noisy, and low-veracity data High and varied volume	EA with fuzzy logic Swarm intelligence, fuzzy-logic-based matching algorithm, EA, and EA

results, classifying each AI approach as NLP, ML, or CI. The second attribute indicates how each approach is impacted by uncertainty, both how the technique itself and the data are uncertain. An overview of the suggested mitigating tactics for dealing with the problems brought on by uncertainty is given in the third column. For instance, the first row of Table 6.2 shows how inadequate training data can cause ambiguity in ML. Using the method of active learning that chooses a subclass of the data is thought to be very important, thus solving the issue of little practice data, is one potential strategy for overcoming this specific uncertainty.

Keep in mind that we went over each big data aspect individually. However, integrating one or more big data qualities will result in an exponential increase in uncertainty, necessitating much more research.

6.7 CHALLENGES AND RECOMMENDATIONS

This work examines how big data analytics and the datasets themselves are affected by uncertainty. The primary objectives are to examine the state of big data analytics methods, how insecurity might adversely affect these approaches, and pinpoint any unresolved concerns. To assist others in the community when creating their procedures, each typical methodology is outlined.

Big data may provide organizations with a wealth of information, but the terabytes or petabytes of data arriving into an organization every day have shown that the structures and architectures in place today are unable to handle the problem. The task of developing technology that can handle all the technological needs of enormous data streams falls to IT scientists. As data volume increases, IT professionals are receiving more calls asking for more ad hoc analysis and compiled reports. If feasible, decision-makers cannot afford to wait hours or days for answers to their questions. Additionally, end users will want tools to access, comprehend, and evaluate these data on their own without having to go back to IT for each request [40]. The seven Vs of big data—velocity, volume, veracity, value, verbosity, versatility, and variety—and the problems they raise are covered in this chapter. Although there has been a lot of study on data volume, diversity, velocity, and truthfulness, there has been far less research on data value, which includes data pertinent to business interests and decision-making in certain fields.

Future field study on several topics has been suggested by this publication. First of all, because they coexist inherently in real-world circumstances, it is crucial to research the relationships between various big data properties. Second, it's important to experimentally assess the scalability and efficiency of current analytics methods when used with large data. Third, to manage real-time decision-making based on enormous volumes of data, in the fields of ML and NLP, new methods and algorithms must be developed. Fourth, there is a need for more effective techniques to describe uncertainty emerging from big data analytics as well as modeling uncertainty in ML and NLP. Additionally, tackling ML issues and uncertainty concerns in data analytics has shown potential for CI techniques. To reduce uncertainty, there aren't enough CI metaheuristics techniques made particularly for large data analytics. Future big data analytics research and development potential in these domains are quite interesting.

Numerous issues are presented in the fields of data science and big data analytics and are strongly related to human applications and processes. A major challenge is managing enormous and constantly growing amounts of data, which calls for sophisticated infrastructure and quick processing techniques. Structured, unstructured, and semi-structured data are all different and varied, which complicates their integration and analysis. To reach valid findings and make wise judgments, it is essential to guarantee the trustworthiness and quality of the data. To foster trust and protect sensitive information, it is crucial to address privacy issues and ensure data security. Deep knowledge of the data and the issue at hand is necessary to choose the best algorithms for a given task. To gain significant insights from complicated datasets, it is crucial to bridge the gap between domain knowledge and technical competence. In dynamic contexts, real-time data analysis and prompt decision-making are becoming more and more crucial. Additionally, as data analytics is influencing a variety of sectors in an ever-increasing way, addressing ethical issues and potential biases in data utilization is essential for providing accurate and impartial insights. Data science and big data analytics may realize their true potential to improve human experiences and spur innovation in a variety of industries by overcoming these obstacles [41].

6.8 CONCLUSION

This chapter has provided an in-depth review of big data, covering and improving the functionality and intelligence of an application. We have also represented the contemporary appeal of big data and data science. Big data is a collection of datasets that are always expanding since data are produced by everyone and for every purpose, including call centers and handheld electronics. The seven critical big data Vs—velocity, verbosity, variety, versatility, volume, veracity, and value—are also discussed in this chapter. The emphasized features will likely provide straightforward and efficient big data management that may be used for value-added applications and research settings. We have now covered the difficulties we encountered within the parameters of our study, as well as potential future areas and lines of inquiry. When processing vast amounts of data, typical pattern analysis methods might be difficult to use. However, uncertainty exists in many different forms, which results in findings of low quality and low accuracy. Overall, we conclude that our investigation into sophisticated analytical and uncertainty solutions based on big data techniques and data science is fruitful and can serve as a model for future work in the field of data science and its practical applications.

REFERENCES

[1] K. U. Jaseena and J. M. David, "Issues, challenges, and solutions: Big data mining," *CS & IT-CSCP*, vol. 4, no. 13, pp. 131–140, 2014.

[2] B. Marr, "How much data do we create every day? The mind-blowing stats everyone should read," *Forbes*, vol. 21, pp. 1–5, 2018.

[3] A. McAfee, E. Brynjolfsson, T. H. Davenport, D. J. Patil, and D. Barton, "Big data: The management revolution," *Harvard Business Review*, vol. 90, no. 10, pp. 60–68, 2012.

[4] A. Lenk, L. Bonorden, A. Hellmanns, N. Roedder, and S. Jaehnichen, "Towards a taxonomy of standards in smart data," In 2015 *IEEE International Conference on Big Data (Big Data)*, Santa Clara, CA, USA, IEEE, 2015, pp. 1749–1754.

[5] S. Hayat, F. Rehman, N. Riaz, H. Sharif, S. Irshad, and S. Shareef, "Using machine learning algorithms to detect dysplasia in Barrett's esophagus," In *2022 2nd International Conference on Digital Futures and Transformative Technologies (ICoDT2)*, Rawalpindi, Pakistan, IEEE, 2022, pp. 1–6.

[6] A. Panimalar, V. Shree, and V. Kathrine, "The 17 V's of big data," *International Research Journal of Engineering and Technology (IRJET)*, vol. 4, no. 9, pp. 3–6, 2017.

[7] N. Sarfraz, F. Rehman, H. Sharif, S. Akram, B. H. Mughal, and A. Ashfaq, "Efficient energy storage systems management in power plants with artificial intelligence and price control," In *2022 3rd International Conference on Innovations in Computer Science & Software Engineering (ICONICS)*, Karachi, Pakistan, IEEE, 2022, pp. 1–5.

[8] M. Chen, S. Mao, and Y. Liu, "Big data: A survey," *Mobile Networks and Applications*, vol. 19, pp. 171–209, 2014.

[9] S. Pouyanfar, Y. Yang, S.-C. Chen, M.-L. Shyu, and S. S. Iyengar, "Multimedia big data analytics: A survey," *ACM Computing Surveys (CSUR)*, vol. 51, no. 1, pp. 1–34, 2018.

[10] X. Wang and Y. He, "Learning from uncertainty for big data: Future analytical challenges and strategies," *IEEE Systems, Man, and Cybernetics Magazine*, vol. 2, no. 2, pp. 26–31, 2016.

[11] I. Manan, F. Rehman, H. Sharif, N. Riaz, M. Atif, and M. Aqeel, "Quantum computing and machine learning algorithms – A review," In *2022 3rd International Conference on Innovations in Computer Science & Software Engineering (ICONICS)*, Karachi, Pakistan, IEEE, 2022, pp. 1–6.

[12] B. Brown et al., *Big Data: The Next Frontier for Innovation, Competition, and Productivity*, McKinsey Global Institute, New York, pp. 1–156, 2011.

[13] R. L. Villars, C. W. Olofson, and M. Eastwood, "Big data: What it is and why you should care," *White Paper, IDC*, vol. 14, pp. 1–14, 2011.

[14] D. Laney, "3D data management: Controlling data volume, velocity and variety," *META Group Research Note*, vol. 6, no. 70, p. 1, 2001.

[15] J. Gantz and D. Reinsel, "Extracting value from chaos," *IDC Iview*, vol. 1142, no. 2011, pp. 1–12, 2011.

[16] A. Jain, *The 5 Vs of Big Data*, IBM Watson Health Perspectives, New York, 2017.

[17] B. Marr, "Really big data at walmart: Real-time insights from their 40+ petabyte data cloud," *Forbes*, vol. 23, p. 1–3, 2017.

[18] D. Saidulu and R. Sasikala, "Machine learning and statistical approaches for big data: Issues, challenges and research directions," *International Journal of Applied Engineering Research*, vol. 12, no. 21, pp. 11691–11699, 2017.

[19] A. Gandomi and M. Haider, "Beyond the hype: Big data concepts, methods, and analytics," *International Journal of Information Management*, vol. 35, no. 2, pp. 137–144, 2015.

[20] H. Xiong, G. Pandey, M. Steinbach, and V. Kumar, "Enhancing data analysis with noise removal," *IEEE transactions on knowledge and data engineering*, vol. 18, no. 3, pp. 304–319, 2006.

[21] M. H. Khalid et al., "A brief overview of deep learning approaches for IoT security," In *2023 4th International Conference on Computing, Mathematics and Engineering Technologies (iCoMET)*, Sukkur, Pakistan, IEEE, 2023, pp. 1–5.

[22] I. B. M. B. Data and A. Hub, "Extracting business value from the 4 V's of big data," *Retrieved July*, vol. 19, no. 2017, pp. 25–60, 2016.

[23] D. Court, "Getting big impact from big data," *McKinsey Quarterly*, vol. 1, no. 1, pp. 52–60, 2015.

[24] F. H. Knight, "Risk, uncertainty, and profit. Library of economics and liberty," *Acceso*, vol. 31, p. 2017, 1921.

[25] C.-W. Tsai, C.-F. Lai, H.-C. Chao, and A. V Vasilakos, "Big data analytics: A survey," *Journal of Big Data*, vol. 2, no. 1, pp. 1–32, 2015.

[26] R. DeLine, "Research opportunities for the big data era of software engineering," In *2015 IEEE/ACM 1st International Workshop on Big Data Software Engineering,* Florence, Italy, IEEE, 2015, pp. 26–29.

[27] V. López, S. Del Río, J. M. Benítez, and F. Herrera, "Cost-sensitive linguistic fuzzy rule based classification systems under the MapReduce framework for imbalanced big data," *Fuzzy Sets and Systems*, vol. 258, pp. 5–38, 2015.

[28] R. H. Hariri, E. M. Fredericks, and K. M. Bowers, "Uncertainty in big data analytics: Survey, opportunities, and challenges," *Journal of Big Data*, vol. 6, no. 1, pp. 1–16, 2019.

[29] A. Skowron, J. Komorowski, Z. Pawlak, and L. Polkowski, "Rough sets perspective on data and knowledge," In Willi Klösgen, Jan M. Zytkow (eds.), *Handbook of Data Mining and Knowledge Discovery,* Oxford University Press, New York, 2002, pp. 134–149.

[30] I. Düntsch and G. Gediga, "Rough set dependency analysis in evaluation studies: An application in the study of repeated heart attacks," *Informatics Research Reports*, vol. 10, pp. 25–30, 1995.

[31] D. G. Brown, "Classification and boundary vagueness in mapping presettlement forest types," *International Journal of Geographical Information Science*, vol. 12, no. 2, pp. 105–129, 1998.

[32] M. A. Mudassir Khan, "Big data analytics evaluation," *International Journal of Engineering Research in Computer Science and Engineering (IJERCSE)*, vol. 5, no. 2, pp. 25–28, 2018.

[33] K. Weiss, T. M. Khoshgoftaar, and D. Wang, "A survey of transfer learning," *Journal of Big Data*, vol. 3, no. 1, pp. 1–40, 2016.

[34] Y. Tsuruoka et al., "Developing a robust part-of-speech tagger for biomedical text," In *Advances in Informatics: 10th Panhellenic Conference on Informatics, PCI 2005*, Volas, Greece, November 11–13, 2005, Proceedings 10, Springer, 2005, pp. 382–392.

[35] J. Fulcher and L. C. Jain, *Computational Intelligence: A Compendium*, vol. 21, Springer, New York, 2008.

[36] A. Gupta, "Big data analysis using computational intelligence and Hadoop: A study," In *2015 2nd International Conference on Computing for Sustainable Global Development (INDIACom)*, New Delhi, India, IEEE, 2015, pp. 1397–1401.

[37] F. Doctor, C.-H. Syue, Y.-X. Liu, J.-S. Shieh, and R. Iqbal, "Type-2 fuzzy sets applied to multivariable self-organizing fuzzy logic controllers for regulating anesthesia," *Applied Soft Computing*, vol. 38, pp. 872–889, 2016.

[38] C. Ma, H. H. Zhang, and X. Wang, "Machine learning for big data analytics in plants," *Trends in Plant Science*, vol. 19, no. 12, pp. 798–808, 2014.

[39] L. Wang, G. Wang, and C. A. Alexander, "Natural language processing systems and big data analytics," *International Journal of Computational Systems Engineering*, vol. 2, no. 2, pp. 76–84, 2015.

[40] A. A. Tole, "Big data challenges," *Database Systems Journal*, vol. 4, no. 3, pp. 31–40, 2013.

[41] R. Agarwal and V. Dhar, "Big data, data science, and analytics: The opportunity and challenge for IS research," *Information Systems Research*, vol. 25, no. 3, pp. 443–448, 2014.

CHAPTER 7

Artificial Intelligence and Machine Learning with Cyber Ethics for the Future World

Wasswa Shafik

Universiti Brunei Darussalam

Dig Connectivity Research Laboratory (DCRLab)

7.1 INTRODUCTION

Machine Learning and Artificial Intelligence (MAI) have emerged as revolutionary technologies shaping the future world. These technologies have the potential to revolutionize industries, improve efficiency, and enhance decision-making processes [1]. AI refers to the simulation of human intelligence in machines, enabling them to perform tasks that typically require human intelligence, such as learning, reasoning, and problem-solving. ML, a subset of AI, focuses on algorithms and statistical models that allow machines to learn and make predictions or decisions without being explicitly programmed [2]. MAI have gained significant traction in recent years across various fields, including healthcare, finance, transportation, and communication. Their applications range from medical diagnosis and personalized recommendations to autonomous vehicles and natural language processing. The rapid advancements in MAI have led to unprecedented opportunities and challenges for individuals, organizations, and society [3].

As MAI technologies become increasingly pervasive, it is crucial to consider the ethical implications associated with their development and deployment. Cyber ethics, a branch of applied ethics, focuses on the moral and ethical issues arising in the context of technology, particularly concerning information security, privacy, and human values [4]. In MAI, cyber ethics is critical in ensuring these technologies are developed and used responsibly, ethically, and in line with societal values. This research explores the intersection of MAI with cyber ethics and sheds light on the importance of ethical considerations in shaping the future world [5]. It examines MAI applications' challenges and dilemmas, such

DOI: 10.1201/9781032648309-9

as algorithmic bias, privacy concerns, transparency, accountability, and fairness [6]. By addressing these ethical concerns, we can strive for an MAI landscape that promotes social good, respects human rights, and minimizes harm.

Moreover, this chapter delves into the existing frameworks, guidelines, and regulations to foster ethical MAI practices. It examines the roles of governments, organizations, and researchers in promoting ethical development and deployment of these technologies [7]. The exploration of future directions and recommendations seeks to provide insights into navigating the ethical challenges that lie ahead. MAI have the potential to revolutionize healthcare by enhancing diagnostics, treatment, and patient care. Machine learning algorithms can analyze vast amounts of medical data to detect patterns, diagnose diseases, and predict patient outcomes [8]. AI-powered tools can automate administrative tasks, improve precision in surgery, and enable personalized medicine.

In the finance industry, MAI are transforming fraud detection, risk assessment, and algorithmic trading processes. ML algorithms can analyze large datasets to identify patterns and anomalies, enhancing fraud detection capabilities. AI-powered chatbots and virtual assistants provide personalized customer service and support. Predictive analytics based on ML models aid in risk assessment and investment decision-making. ML algorithms enable vehicles to perceive their surroundings, make real-time decisions, and navigate safely [9]. AI-based traffic management systems optimize traffic flow, reduce congestion, and enhance transportation efficiency. Also, ML algorithms are employed in logistics and supply chain management to optimize routing and improve delivery processes, as depicted in Figure 7.1.

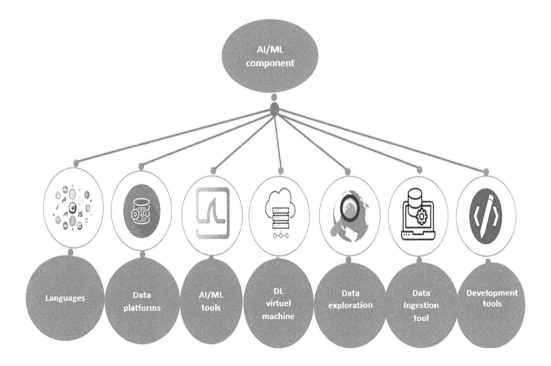

FIGURE 7.1   Artificial intelligence/machine learning components.

Adaptive learning platforms powered by ML algorithms can assess students' strengths and weaknesses, tailoring educational content accordingly. Intelligent tutoring systems provide personalized guidance and support. Natural language processing capabilities enable automated essay grading and language learning assistance [10]. Recommendation systems analyze user preferences, purchase history, and browsing behavior to provide personalized product suggestions. ML algorithms help optimize inventory management, demand forecasting, and pricing strategies. Chatbots and virtual assistants enhance customer service and support. ML algorithms analyze sensor data to detect anomalies, predict equipment failures, and optimize maintenance schedules [11]. AI-powered robotics and automation systems enhance productivity and efficiency. ML-based quality control systems detect defects and optimize production parameters.

7.1.1 The Contribution of this Chapter

The following are the contributions of this chapter.

- Emphasizes the significance of ethical considerations in developing and deploying MAI technologies.

- Highlights the need to address biases and ensure fairness in algorithmic decision-making, stressing the importance of protecting privacy and implementing robust data governance practices.

- Highlights the importance of maintaining human oversight and control over these technologies, promoting collaboration and multidisciplinary approaches in addressing ethical challenges.

- Underlines the importance of continuous monitoring, evaluation, and adaptation in ethical MAI development, advocating for public engagement and inclusion in MAI development processes.

- Emphasizes the need for regulatory frameworks and international collaboration to establish clear ethical standards.

- Encourages education and awareness initiatives to foster a responsible MAI development culture.

7.1.2 The Organization of this Chapter

The remainder of this chapter is divided into seven sections as follows: Section 7.2 presents an understanding of artificial intelligence and machine learning. Section 7.3 demonstrated the applications of artificial intelligence and machine learning. Section 7.4 presents cyber ethics and its importance in the future world. Section 7.5 presents the ethical challenges in artificial intelligence and machine learning. Section 7.6 demonstrates ensuring ethical MAI development. Section 7.7 illustrates future directions, lessons learned, recommendations, and the conclusion of this chapter.

7.2 UNDERSTANDING MACHINE LEARNING AND ARTIFICIAL INTELLIGENCE

AI refers to developing computer systems or machines that can perform tasks that typically require human intelligence. It involves the simulation of human cognitive processes, such as learning, reasoning, problem-solving, perception, and language understanding, in machines [2]. ML, a subset of AI, focuses on developing algorithms and statistical models that enable machines to learn from data and make predictions or decisions without being explicitly programmed. Instead of following predefined rules, ML algorithms learn patterns and relationships from large datasets, enabling them to generalize and make accurate predictions or decisions on new, unseen data [4].

The development of MAI technologies has progressed significantly over the years. Initially, AI was primarily focused on rule-based systems that relied on explicit programming to solve specific problems. However, the advent of ML algorithms, particularly in the past decade, has revolutionized AI. ML algorithms can automatically learn and improve from data, making them more adaptable and capable of handling complex tasks. Currently, MAI technologies are in a state of rapid advancement [3]. ML algorithms, such as deep learning, have demonstrated remarkable success in various domains, including computer vision, natural language processing, and speech recognition [12]. The availability of vast amounts of data and increased computing power has facilitated the training of more complex models, enabling breakthroughs in areas like image and speech recognition, autonomous vehicles, and language translation [13]. The benefits of MAI adoption are numerous, as summarized.

- **Automation and Efficiency:** MAI technologies can automate repetitive and mundane tasks, increasing efficiency and productivity [14]. This allows human workers to focus on more complex and creative endeavors.

- **Data-Driven Decision-Making:** ML algorithms can analyze large volumes of data and extract valuable insights, enabling data-driven decision-making and improving the accuracy of predictions and recommendations [12].

- **Personalization:** MAI algorithms can personalize experiences, such as personalized product recommendations in e-commerce, personalized learning in education, or personalized healthcare treatments [15].

- **Improved Customer Service:** AI-powered chatbots and virtual assistants can provide immediate and personalized customer support, improving customer satisfaction and reducing response times [16].

However, along with the benefits, there are also challenges associated with MAI adoption:

- **Ethical Considerations:** MAI technologies raise ethical concerns, such as privacy, bias, and transparency. It is crucial to ensure that these technologies are developed and deployed in a manner that respects human values, fairness, and accountability [17].

- **Job Displacement:** Automating certain tasks through MAI may lead to job displacement or require individuals to acquire new skills to adapt to changing job requirements [18].

- **Data Bias and Quality:** ML algorithms heavily rely on data, and biases or poor data quality can lead to biased or inaccurate outcomes [5]. Care must be taken to ensure diverse and representative datasets.

- **Security and Privacy:** MAI technologies raise data security and privacy concerns. Safeguarding sensitive information and preventing unauthorized access is crucial [18].

Adopting MAI brings numerous benefits, including automation, data-driven decision-making, personalization, and improved customer service [15]. However, ethical considerations, job displacement, data bias and quality, and security and privacy must be carefully addressed to ensure MAI technologies' responsible and beneficial deployment.

7.3 APPLICATIONS OF MACHINE LEARNING AND ARTIFICIAL INTELLIGENCE

These applications represent just a fraction of the wide-ranging impact of ML and AI. As technology advances, ML and AI's potential to transform industries and enhance various aspects of our lives will only continue to grow.

7.3.1 Healthcare

ML and AI play a decisive role in healthcare. They are used in medical imaging analysis to detect and diagnose diseases such as cancer. ML algorithms can analyze large volumes of medical images, such as X-rays and MRI scans, to identify patterns and anomalies [19]. AI-powered systems assist in treatment planning, predicting patient outcomes, and personalized medicine. Additionally, MAI facilitate drug discovery by analyzing vast datasets and identifying potential candidates for new medications.

AI algorithms can scrutinize patient data and recognize potential health issues, allowing for treatment and early intervention. Moreover, AI has the potential to revolutionize drug discovery and development, enlightening the efficiency and speed of the process [18]. In addition to its benefits in specific industries, AI has become essential for data analysis and management in the digital era. As more data is generated and collected, AI algorithms can analyze and identify patterns, insights, and trends, helping organizations make data-driven decisions [20]. Moreover, AI can improve data security by detecting and mitigating cyber threats; some applications are presented in Figure 7.2 and the way it fosters manufacturing.

7.3.2 Finance

In the finance industry, ML algorithms are employed in fraud detection systems. By analyzing patterns and anomalies in transaction data, AI can identify fraudulent activities and alert financial institutions. AI-powered chatbots and virtual assistants provide

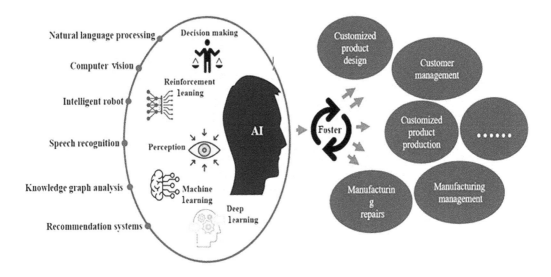

FIGURE 7.2   Application of artificial intelligence and its customized manufacturing.

personalized customer support and enhance user experiences [21]. Furthermore, ML algorithms analyze financial data and market trends to predict stock prices, optimize investment strategies, and perform algorithmic trading.

7.3.3 Transportation

MAI technologies are transforming transportation, particularly in developing autonomous vehicles. ML algorithms enable vehicles to perceive their surroundings through sensors and make real-time decisions based on the collected data [17]. This technology is crucial for safe navigation and collision avoidance. Additionally, AI-powered systems optimize traffic flow by analyzing historical and real-time data, predicting congestion patterns, and suggesting alternate routes [10]. Logistic companies utilize AI to optimize routing and delivery processes, reducing costs and improving efficiency.

7.3.4 Natural Language Processing (NLP)

NLP focuses on enabling machines to understand, interpret, and generate human language. ML and AI techniques are applied in various NLP applications. Chatbots and virtual assistants utilize NLP to provide conversational interfaces and assist users with tasks such as answering questions, making recommendations, and performing transactions [12]. Voice recognition systems leverage NLP to convert spoken language into written text, enabling hands-free operation and voice commands.

7.3.5 Image and Speech Recognition

ML algorithms excel in image and speech recognition tasks. Image recognition applications include facial recognition systems used for security and identity verification. ML algorithms can also detect objects in images, classify images into various categories, and assist in autonomous driving by identifying road signs and pedestrians [22]. Speech recognition technology enables voice assistants, transcription services, and speech-to-text applications, enhancing accessibility and enabling hands-free interaction with devices.

7.3.6 Manufacturing and Quality Control

ML algorithms are utilized in the manufacturing industry to optimize processes and improve quality control. Predictive maintenance is a prominent application where ML algorithms analyze sensor data to detect patterns indicative of potential equipment failures [12]. Maintenance can be scheduled proactively by identifying issues before they occur, reducing downtime and optimizing productivity. ML algorithms are also employed in quality control processes, detecting defects in real time, ensuring product consistency, and reducing waste.

7.3.7 Recommender Systems and E-commerce

These systems analyze user preferences, browsing behavior, and purchase history to provide personalized product recommendations. By suggesting relevant items, recommendation systems enhance the user experience, increase customer engagement, and drive sales [23]. ML algorithms also contribute to dynamic pricing strategies, allowing businesses to optimize pricing based on factors like demand, competition, and customer behavior.

7.3.8 Cybersecurity

ML and AI are instrumental in combating cyber threats. ML algorithms analyze network traffic and user behavior, identifying patterns of malicious activity and anomalies that could indicate a potential attack [24]. By continuously learning from new data, these algorithms can detect and prevent various cyber threats, such as malware, phishing attacks, and network intrusions. AI systems also help identify vulnerabilities in systems and networks, enhancing overall cybersecurity.

7.3.9 Energy and Sustainability

ML and AI technologies contribute to energy optimization and sustainability efforts. ML algorithms analyze energy consumption patterns, predict demand, and optimize energy distribution in smart grids. By dynamically managing energy usage, AI systems can reduce waste and increase efficiency [25]. ML algorithms also facilitate environmental monitoring by analyzing satellite imagery, weather data, and sensor data to assess air quality, monitor wildlife populations, and support environmental conservation efforts.

7.3.10 Education and Personalized Learning

ML algorithms are used in personalized learning platforms in education. These platforms adapt to individual student's learning styles, preferences, and progress, providing customized educational content and activities. By analyzing data on student performance, ML algorithms can identify areas where additional support is needed and recommend appropriate resources or interventions [21]. Personalized learning systems enhance engagement, improve learning outcomes, and support individualized instruction. The following section presents some cyber ethics and its merits.

7.4 CYBER ETHICS AND ITS IMPORTANCE IN THE FUTURE WORLD

Cyber ethics refers to the ethical principles and guidelines that govern the use of technology, particularly in the digital realm. Cyber ethics has become increasingly important in shaping the future world with the rapid advancement of technology, including MAI. The following are some of the explored significance of cyber ethics and its implications for our digital society.

7.4.1 Significance of Cyber Ethics

By integrating ethical principles into developing, deploying, and using these technologies, we can harness their potential while upholding values such as fairness, transparency, privacy, and accountability [26]. Cyber ethics provides the necessary framework to navigate the complex ethical challenges in our increasingly digital and interconnected world, shaping a future that benefits individuals, society, and the global community.

7.4.1.1 Ethical Decision-Making

Cyber ethics plays a vital role in ensuring ethical decision-making processes in MAI. As these technologies become more autonomous and capable of making decisions that impact individuals and society, it is essential to consider the ethical implications [6]. Cyber ethics guides the development and deployment of MAI systems, emphasizing the importance of fairness, transparency, accountability, and respect for human values.

7.4.1.2 Privacy and Data Protection

Privacy is a fundamental concern in the digital age. MAI technologies rely on vast amounts of data to function effectively. Cyber ethics ensures the responsible handling of personal data, protecting individuals' privacy rights [23]. It emphasizes the need for informed consent, secure data storage, and data anonymization to mitigate the risk of data breaches and unauthorized access.

7.4.1.3 Bias and Fairness

MAI algorithms can inadvertently perpetuate biases in the data they are trained on. Cyber ethics calls for fairness in algorithmic decision-making and identifying and mitigating biases [27]. It encourages diverse and representative datasets and the continual monitoring and assessment of AI systems for fairness and equity.

7.4.1.4 Explainability and Transparency

MAI models often operate as "black boxes," making understanding how they arrive at their decisions challenging. Cyber ethics advocates for transparency and explainability in AI systems, enabling users and stakeholders to understand the reasoning behind the outcomes [26]. This promotes accountability, trust, and responsible deployment of MAI technologies.

7.4.1.5 Accountability and Liability

Cyber ethics addresses the issue of accountability and liability when MAI systems cause harm or make erroneous decisions. It emphasizes the need for clear lines of responsibility, ensuring that individuals or entities are held accountable for the actions and consequences of MAI technologies [19]. This includes addressing potential legal and ethical implications, such as liability for autonomous vehicle accidents.

7.4.1.6 Economic and Social Impact

MAI have the potential to impact society and the economy in significant ways. Cyber ethics ensures that these technologies are developed and used to benefit society, uphold human rights, and minimize negative consequences [28]. It calls for considerations of the societal impact of MAI, including job displacement, economic inequality, and the distribution of benefits and risks.

7.4.1.7 Global Governance and Collaboration

Cyber ethics highlights the need for global collaboration and governance frameworks to address ethical challenges in MAI. It encourages the involvement of multiple stakeholders, including governments, organizations, researchers, and individuals, in shaping policies, guidelines, and regulations [29]. Collaboration fosters a collective effort to ensure ethical MAI development and promotes a globally harmonized approach to cyber ethics.

7.4.2 Potential Risks and Negative Consequences

The rapid advancement and widespread adoption of ML and AI technologies without appropriate ethical guidelines can give rise to several risks and negative consequences. At this point, we present some potential implications:

7.4.2.1 Bias and Discrimination

MAI systems can inadvertently perpetuate and amplify biases in the data they are trained on. These biases can result in discriminatory outcomes in hiring, lending, and law enforcement without ethical guidelines [30]. Unchecked MAI systems can perpetuate societal biases, reinforce inequalities, and lead to unfair treatment of individuals or marginalized groups.

7.4.2.2 Data Misuse and Privacy

Without robust ethical guidelines, there is a risk of privacy breaches and data misuse. MAI systems often rely on vast amounts of personal data, and inappropriate handling can compromise individuals' privacy rights. A lack of ethical guidelines can lead to unauthorized data collection, surveillance, or the use of personal data for unintended purposes, potentially resulting in privacy violations and breaches of trust [31].

7.4.2.3 Accountability and Lack of Transparency

Ethical guidelines promote transparency and accountability in MAI systems. Without such guidelines, systems can operate as "black boxes," making it difficult to understand the decision-making process and assess the accountability of these technologies [32]. This lack of transparency can erode trust, hinder effective governance, and impede addressing potential biases, errors, or adverse outcomes.

7.4.2.4 Job Displacement and Unemployment

MAI technologies have the potential to automate various tasks, which can lead to job displacement and unemployment. Without ethical guidelines, the rapid adoption of these technologies without adequate support for affected workers can exacerbate economic inequalities and social challenges [5]. Ethical considerations should include measures for reskilling and upskilling the workforce to mitigate the negative impact on employment.

7.4.2.5 Safety and Security Risks

MAI systems that lack ethical guidelines can pose safety and security risks. For instance, autonomous vehicles without appropriate ethical guidelines and safety standards may endanger lives. Inadequate security measures can make MAI systems vulnerable to malicious attacks and manipulation, leading to significant consequences in sectors like healthcare, transportation, and critical infrastructure [15].

7.4.2.6 Decision-Making and Ethical Dilemmas

MAI technologies can face ethical dilemmas where they must make complex moral decisions. Without ethical guidelines, these systems may be unable to make morally sound decisions, potentially leading to unintended consequences or ethical conflicts [19].

7.4.2.7 Socioeconomic Implications

Deploying MAI technologies without ethical guidelines can exacerbate socioeconomic disparities. If these technologies are primarily accessible to privileged groups or organizations, they can deepen inequalities and widen the digital divide. Ethical guidelines should address equitable access, fairness, and the broader societal implications of MAI deployment [12]. To mitigate these risks and negative consequences, it is essential to establish comprehensive ethical guidelines and frameworks that guide the development, deployment, and use of MAI technologies. These guidelines should prioritize fairness, transparency, accountability, privacy protection, and the consideration of societal impact [33]. By incorporating ethical principles, we can harness the potential of MAI while minimizing their negative consequences and ensuring their alignment with human values and societal well-being.

7.5 ETHICAL CHALLENGES IN MACHINE LEARNING AND ARTIFICIAL INTELLIGENCE

Ethical considerations must be integrated into the entire lifecycle of ML systems, from data collection and model development to deployment and ongoing monitoring. By proactively addressing these ethical concerns, we can maximize the benefits of ML while minimizing potential harms and building a responsible and inclusive technological future.

7.5.1 Bias and Fairness

ML algorithms learn from historical data, which can inadvertently reflect societal biases and discrimination. ML models can perpetuate and amplify biases in decision-making processes without careful consideration. For instance, in hiring or loan approval systems, biased algorithms can discriminate against certain groups based on gender, race, or socioeconomic background [18]. Ensuring fairness in ML systems requires actively identifying and addressing biases in training data and implementing measures to mitigate unfair outcomes.

7.5.2 Data Protection

ML often relies on large amounts of data, including personal and sensitive information. Ethical concerns arise around collecting, storing, and using these data. Protecting individuals' privacy rights is essential. This includes obtaining informed consent, implementing strong security measures to prevent data breaches, and ensuring data is used only for intended purposes [9]. Anonymization techniques can also be applied to protect individuals' identities while still enabling effective ML training.

7.5.3 Explainability and Transparency

ML algorithms often operate as "black boxes," making understanding how they arrive at their decisions difficult. This lack of transparency can erode trust and raise ethical concerns, especially when ML systems are deployed in critical healthcare or criminal justice domains [10]. Ethical guidelines should promote transparency and explainability, allowing individuals to understand and challenge algorithmic decisions. Techniques such as interpretable ML models and model-agnostic explanations can provide meaningful insights into ML decision-making processes.

7.5.4 Human Control and Autonomy

ML systems are becoming more autonomous, with the ability to make decisions without direct human intervention. Ethical questions arise around the balance of control between humans and ML systems. Ensuring that humans retain the ability to override or intervene in ML-generated decisions is crucial, particularly in domains where human values, ethics, and context-specific knowledge are essential [11]. Striking the right balance of human control and system autonomy is a key ethical consideration.

7.5.5 Algorithmic Accountability

ML algorithms can significantly impact individuals and society, but responsibility for their decisions and consequences may be unclear. Ethical guidelines should establish mechanisms for algorithmic accountability [18]. This includes auditing, monitoring, and assessing the impact of ML systems and providing avenues for recourse in cases of errors, biases, or unfair outcomes. Accountability ensures that developers and organizations bear responsibility for the effects of their ML systems.

7.5.6 Security and Safety

ML systems can pose safety and security risks in applications like autonomous vehicles or healthcare; the reliability and safety of ML algorithms are crucial to prevent harm to individuals. In addition, ML systems may be vulnerable to adversarial attacks, where intentionally manipulated inputs can cause the algorithms to make incorrect or malicious decisions [21]. Ethical considerations involve developing robust security measures, including safeguards against adversarial attacks, to ensure the integrity and safety of ML systems.

7.5.7 Economic Implications

The widespread adoption of ML technologies can have socioeconomic implications, including job displacement, economic inequality, and the concentration of power. Ethical guidelines should address these implications by promoting policies that support affected individuals, such as reskilling and upskilling programs [34]. Additionally, considerations for equitable access to ML benefits and addressing disparities in ML deployment are essential to prevent exacerbating existing socioeconomic inequalities.

7.5.8 Environmental Impact

ML models often require significant computing power and energy consumption, contributing to environmental impact. Ethical guidelines should encourage the development of energy-efficient algorithms, optimization techniques, and responsible use of computing resources to minimize the environmental footprint of ML [32]. Addressing these ethical impacts requires interdisciplinary collaboration among researchers, policymakers, industry stakeholders, and the public. Ethical considerations must be integrated into the entire lifecycle of ML systems, from data collection and model development to deployment and ongoing monitoring.

7.6 ENSURING ETHICAL MAI DEVELOPMENT

Ensuring the ethical development of MAI involves adopting practices and guidelines prioritizing fairness, transparency, accountability, privacy, and promoting human well-being.

7.6.1 Ethical Frameworks and Guidelines

Establishing ethical frameworks and guidelines is essential to provide a foundation for responsible MAI development. These frameworks should encompass fairness, transparency, accountability, privacy, and human rights [16]. They should guide the design, development, deployment, and use of MAI systems, addressing these technologies' potential risks and negative consequences.

7.6.2 Data Governance and Bias Mitigation

Addressing biases in data is crucial to ensure fairness in MAI systems. Robust data governance practices should be implemented to identify and mitigate biases in training datasets [33]. Data collection should be diverse, representative, and inclusive to avoid perpetuating societal biases. Regular audits and assessments of datasets should be conducted to identify and rectify potential biases throughout the development process.

7.6.3 Transparency and Explainability

Enhancing transparency and explainability in MAI systems is essential for building trust and accountability. ML models and algorithms should be designed to provide interpretable outputs, enabling users to understand how decisions are made [35]. Employing techniques such as model interpretability, algorithmic explanations, and visualizations can help stakeholders comprehend the factors contributing to MAI outcomes.

7.6.4 Privacy and Data Protection

Privacy protection is a critical ethical consideration in MAI development. Adequate measures should be implemented to safeguard personal data and ensure compliance with relevant privacy regulations. Anonymization techniques, data minimization, secure storage, and informed consent practices should be followed to protect individuals' privacy rights and prevent unauthorized access or misuse of data [36].

7.6.5 Human Oversight and Control

MAI systems should incorporate mechanisms for human oversight and control to prevent undue concentration of power and ensure accountability. Humans should be able to understand, question, and override system decisions when necessary [37]. Human involvement is crucial in high-stakes domains like healthcare, law enforcement, and critical infrastructure, where human values, ethics, and context-specific knowledge are indispensable.

7.6.6 Ethical Review and Impact Assessment

Prioritizing ethical review and impact assessments of MAI systems can help identify potential risks, biases, and unintended consequences. Independent ethical review boards or committees can assess MAI projects' societal, ethical, and human rights implications [27]. Regular audits and assessments should be conducted to evaluate the system's performance, fairness, safety, and adherence to ethical guidelines.

7.6.7 Collaboration and Multidisciplinary Approaches

Ethical MAI development requires collaboration among diverse stakeholders, including researchers, policymakers, industry experts, ethicists, and representatives from affected communities. Multidisciplinary approaches can ensure a holistic consideration of ethical implications and diverse perspectives [22]. Collaboration can lead to the development of comprehensive frameworks, guidelines, and regulatory measures that promote responsible MAI practices.

7.6.8 Continuous Monitoring and Adaptation

Ethical MAI development is an ongoing process that requires continuous monitoring and adaptation. Regular assessments should be conducted to identify emerging ethical challenges, update guidelines, and adapt to evolving societal needs and values. Continuous monitoring can help identify biases, unintended consequences, or potential harms, enabling timely interventions and adjustments to ensure ethical compliance [38]. By incorporating these considerations into the development process, stakeholders can work towards creating

MAI systems that are fair, transparent, accountable, and aligned with human values. The collaborative effort of researchers, policymakers, industry leaders, and the public is crucial to ensure that MAI technologies are developed and deployed responsibly, promoting societal well-being, and minimizing potential ethical risks.

7.7 FUTURE DIRECTIONS, LESSONS LEARNED, AND RECOMMENDATIONS

These efforts can contribute to developing and deploying MAI technologies that align with human values, respect individual rights, and promote societal well-being in the future.

7.7.1 Future Directions and Recommendations

7.7.1.1 Ethical Standards and Regulations

Governments and regulatory bodies should collaborate to establish ethical standards and regulations for MAI development. These standards should encompass fairness, transparency, accountability, and privacy. Legal frameworks can help enforce compliance and provide consequences for unethical practices [19].

7.7.1.2 Education and Awareness

Increasing education and awareness about ethical MAI practices is crucial. This includes educating developers, researchers, policymakers, and the public about the ethical implications, challenges, and best practices in MAI [39]. Ethical training programs and resources can help foster a culture of responsible MAI development.

7.7.1.3 Interdisciplinary Collaboration

Encouraging interdisciplinary collaboration is essential for addressing the ethical dimensions of MAI. Collaboration between experts in technology, ethics, law, social sciences, and other relevant fields can ensure a holistic consideration of ethical implications and diverse perspectives [26]. This collaboration can lead to more comprehensive ethical guidelines and frameworks.

7.7.1.4 Ethical Review Boards

Establishing independent ethical review boards or committees can help assess the ethical implications of MAI projects. These boards can conduct comprehensive reviews, evaluate potential risks, biases, and societal impact, and provide recommendations to ensure ethical compliance [25]. The involvement of diverse stakeholders in these boards is crucial to ensure transparency and accountability.

7.7.1.5 International Collaboration and Governance

MAI development transcends national boundaries. International collaboration is essential to establish common ethical guidelines, share best practices, and address global challenges. International organizations and partnerships can be vital in facilitating discussions, coordinating efforts, and promoting responsible MAI development worldwide [20].

7.7.1.6 Public Engagement and Inclusion

The involvement of the public in shaping MAI policies and practices is crucial. Encouraging public participation through consultations, deliberative processes, and citizen juries can ensure that MAI technologies align with societal values and address the concerns of diverse communities [29]. This inclusivity promotes accountability and fosters trust in MAI systems.

7.7.1.7 Auditing and Certification

Developing mechanisms for auditing and certifying the ethical compliance of MAI systems can enhance accountability and trust. Independent auditing bodies can assess MAI algorithms, data handling practices, and adherence to ethical guidelines [40]. Certification programs can help users and stakeholders identify trustworthy and ethically developed MAI solutions.

7.7.1.8 Responsible Data Sharing and Collaboration

Encouraging responsible data sharing and collaboration can promote ethical MAI development. This includes promoting open datasets, ensuring data privacy, and establishing guidelines for data sharing between organizations [41]. Collaboration platforms that facilitate sharing best practices, ethical challenges, and lessons learned can foster responsible development.

7.7.1.9 Ethical Considerations in Research Funding

Research funding agencies should incorporate ethical considerations into their evaluation criteria. Supporting research projects prioritizing ethical MAI development can incentivize responsible practices and contribute to advancing ethically aligned technologies [13].

7.7.1.10 Continuous Monitoring and Evaluation

Continuous monitoring, evaluation, and adaptation are essential for ensuring ongoing ethical MAI development. Regular assessments should be conducted to identify emerging ethical challenges, update guidelines, and adapt to changing societal needs and values [42]. Ongoing feedback loops and mechanisms for user feedback can help improve ethical practices over time.

7.7.2 Lessons Learned from this Chapter

Within this subsection, lessons learned from this chapter are demonstrated followed by the conclusion.

7.7.2.1 Ethical Considerations Are Crucial

This chapter emphasizes that ethical considerations are paramount in developing and deploying MAI technologies. The potential impact of these technologies on individuals, society, and the environment necessitates a strong focus on principles for example, fairness, transparency, privacy, accountability, and the promotion of human well-being [17]. Ethical guidelines and frameworks should be integrated into the entire MAI systems lifecycle.

7.7.2.2 Bias Awareness and Mitigation

This chapter highlights the need to be aware of biases in MAI systems. ML algorithms learn from historical data, which can reflect societal biases and discrimination. Recognizing and addressing these biases is essential to ensure fairness in algorithmic decision-making [43]. Mitigation strategies, such as diverse and representative training data, algorithmic adjustments, and ongoing monitoring, should be employed to avoid perpetuating biases and discriminatory outcomes.

7.7.2.3 Privacy Protection and Data Governance

The collection, storage, and use of personal data in MAI systems raise significant privacy concerns. To address this, robust data governance practices should be implemented. Informed consent should be obtained from individuals, and secure data handling protocols should be followed to protect data from unauthorized access or misuse [31]. Techniques such as data anonymization can help protect privacy while enabling effective ML training.

7.7.2.4 Transparency and Explainability

This chapter emphasizes the importance of transparency and explainability in MAI systems. AI algorithms often operate as "black boxes," making it difficult for users and stakeholders to understand the reasoning behind their decisions. Ensuring transparency and explainability builds trust, enables accountability, and allows individuals to understand and challenge algorithmic outcomes [12]. Techniques such as model interpretability, algorithmic explanations, and visualizations can provide meaningful insights into ML decision-making processes.

7.7.2.5 Human Oversight and Control

This chapter highlights the need for human control and oversight over MAI systems. While autonomy is desirable for these technologies, striking a balance between system autonomy and human intervention is crucial. Human involvement in decision-making processes remains essential in domains where human values, ethics, and contextual knowledge are indispensable [14]. This ensures that MAI systems align with societal values and address complex moral and ethical considerations.

7.7.2.6 Collaboration and Multidisciplinary Approaches

This chapter emphasizes the importance of collaboration and multidisciplinary approaches in addressing the ethical challenges of MAI. Combining researchers, policymakers, industry experts, ethicists, and the public allows for a comprehensive consideration of ethical implications and diverse perspectives [44]. Collaboration ensures that ethical guidelines and frameworks encompass various dimensions and contribute to responsible MAI practices.

7.7.2.7 Continuous Monitoring and Adaptation

This chapter underscores the importance of continuous monitoring, evaluation, and adaptation in ethical MAI development. Ongoing assessments, audits, and impact evaluations help identify emerging ethical challenges, update guidelines, and address biases or unintended consequences [15]. Ethical considerations should be integrated into regular evaluations to ensure the continuous improvement of MAI systems and practices.

7.7.2.8 Public Engagement and Inclusion

This chapter highlights the significance of public engagement and inclusion in MAI development processes. Engaging the public through consultations, deliberative processes, and citizen juries fosters accountability, transparency, and trust [23]. It ensures that MAI technologies align with societal values, address concerns of diverse communities, and mitigate potential risks or negative impacts.

7.7.2.9 Regulatory Frameworks and International Collaboration

This chapter stresses the importance of establishing regulatory frameworks and ethical standards for MAI development. Governments and regulatory bodies should collaborate to create clear guidelines and regulations that promote ethical practices [24]. International collaboration and governance frameworks are crucial to ensure consistent ethical guidelines, share best practices, and address global challenges associated with MAI.

7.7.2.10 Education and Awareness

This chapter emphasizes the need for education and awareness about ethical MAI practices. Building a responsible MAI development culture requires educating developers, researchers, policymakers, and the public about these technologies' ethical implications, challenges, and best practices [45]. Educational initiatives, training programs, and resources should be provided to foster a deep understanding of ethical considerations in MAI.

7.8 CONCLUSION

This rapid advancement of these technologies calls for responsible practices that prioritize fairness, transparency, accountability, privacy, and the promotion of human well-being. This chapter underscores the need to address biases and ensure fairness in algorithmic decision-making. It emphasizes protecting privacy and implementing robust data governance practices to safeguard personal data. Transparency and explainability in MAI systems are essential for building trust, enabling accountability, and understanding algorithmic decisions. Maintaining human oversight and control over these technologies strikes a balance between system autonomy and human intervention. Collaboration and multidisciplinary approaches are vital for addressing the ethical challenges associated with MAI. Comprehensive ethical guidelines and frameworks can be developed by involving researchers, policymakers, industry experts, ethicists, and the public. Continuous monitoring, evaluation, and adaptation are necessary to identify emerging ethical challenges and ensure the ongoing improvement of MAI systems. Public engagement and inclusion play a crucial role in ensuring that MAI technologies align with societal values and address the concerns of diverse communities. Regulatory frameworks and international collaboration are necessary to establish clear ethical standards and promote responsible practices globally. Education and awareness initiatives are essential for fostering a responsible

MAI development culture and ensuring a deep understanding of ethical implications. By embracing these lessons learned and implementing the recommendations provided, stakeholders can work towards the responsible development and deployment of MAI technologies. By prioritizing ethical considerations, we can harness the potential of these technologies while mitigating risks, protecting individual rights, and promoting a future world where MAI contribute positively to society and human well-being.

REFERENCES

[1] T. Mazhar, H. M. Irfan, I. Haq, I. Ullah, M. Ashraf, et al., "Analysis of challenges and solutions of IoT in smart grids using AI and machine learning techniques: A review," *Electronics*, 12(1):242, 2023, https://doi.org/10.3390/electronics12010242.

[2] I. H. Sarker, A. I. Khan, Y. B. Abushark and F. Alsolami, "Internet of things (iot) security intelligence: A comprehensive overview, machine learning solutions and research directions," *Mobile Networks and Applications*, 28(1):296–312, 2023, https://doi.org/10.1007/s11036-022-01937-3.

[3] W. Shafik, "IoT-based energy harvesting and future research trends in wireless sensor networks," In K. Hemant Kumar Reddy, Diptendu Sinha Roy, Tapas Kumar Mishra, Mir Wajahat Hussain (eds.). *Handbook of Research on Network-Enabled IoT Applications for Smart City Services*, IGI Global, pp. 282–306, 2023, https://doi.org/10.4018/979-8-3693-0744-1.ch016.

[4] A. A. Khan, A. A. Laghari, P. Li, M. A. Dootio and S. Karim, "The collaborative role of blockchain, artificial intelligence, and industrial internet of things in digitalization of small and medium-size enterprises," *Scientific Reports*, 13(1):1656, 2023, https://doi.org/10.1038/s41598-023-28707-9.

[5] W. Shafik, "Making cities smarter: IoT and SDN applications, challenges, and future trends," In Poshan Yu, Xiaohan Hu, Ajai Prakash, Nyaribo Wycliffe Misuko, and Gu Haiyue (eds.), *Opportunities and Challenges of Industrial IoT in 5G and 6G Networks*, pp. 73–94, 2023, IGI Global, https://doi.org/10.4018/978-1-7998-9266-3.ch004.

[6] A. Mitra, B. Bera, A. K. Das, S. S. Jamal and I. You, "Impact on blockchain-based AI/ML-enabled big data analytics for Cognitive Internet of Things environment," *Computer Communications*, 197:173–185, 2023, https://doi.org/10.1016/j.comcom.2022.10.010.

[7] C. Greco, G. Fortino, B. Crispo and K. K. Choo, "AI-enabled IoT penetration testing: State-of-the-art and research challenges," *Enterprise Information Systems*, 17(9):2130014, 2023, https://doi.org/10.1080/17517575.2022.2130014.

[8] T. M. Ghazal, M. K. Hasan, M. Ahmad, H. M. Alzoubi and M. Alshurideh, "Machine learning approaches for sustainable cities using internet of things," In Muhammad Alshurideh, Barween Hikmat Al Kurdi , Ra'ed Masa'deh, Haitham M. Alzoubi, and Said Salloum (eds.). *The Effect of Information Technology on Business and Marketing Intelligence Systems*, pp. 1969–1986, 2023, Cham: Springer International Publishing, https://doi.org/10.1007/978-3-031-12382-5_108.

[9] W. Shafik, "A comprehensive cybersecurity framework for present and future global information technology organizations," In Festus Fatai Adedoyin and Bryan Christiansen (eds.). *Effective Cybersecurity Operations for Enterprise-Wide Systems*, pp. 56–79, 2023, IGI Global, https://doi.org/10.4018/978-1-6684-9018-1.ch002.

[10] M. Chandra, K. Kumar, P. Thakur, S. Chattopadhyaya, F. Alam and S. Kumar, "Digital technologies, healthcare and Covid-19: Insights from developing and emerging nations," *Health and Technology*, 12(2):547–568, 2022, https://doi.org/10.1007/s12553-022-00650-1.

[11] H. Wang, K. Qin, R.Y. Zakari, G. Lu and J. Yin, "Deep neural network-based relation extraction: An overview," *Neural Computing and Applications*, 34:1–21, 2022, https://doi.org/10.1007/s00521-021-06667-3.

[12] P. Nilsen, J. Reed, M. Nair, C. Savage, C. Macrae, et al., "Realizing the potential of artificial intelligence in healthcare: learning from intervention, innovation, implementation and improvement sciences," *Frontiers in Health Services*, 2:961475, 2022, https://doi.org/10.3389/frhs.2022.961475.

[13] E. Tokgöz and M. A. Carro, "Applications of artificial intelligence, machine learning, and deep learning on facial plastic surgeries," In Emre Tokgöz and Marina A. Carro (eds.). *Cosmetic and Reconstructive Facial Plastic Surgery: A Review of Medical and Biomedical Engineering and Science Concepts*, pp. 281–306, 2023, Cham: Springer Nature Switzerland, https://doi.org/10.1007/978-3-031-31168-0_9.

[14] S. N. Alaziz, B. Albayati, A. A. El-Bagoury and W. Shafik, "Clustering of covid-19 multi-time series-based k-means and PCA with forecasting," *International Journal of Data Warehousing and Mining*, 19(3):1–25, 2023, https://doi.org/10.4018/IJDWM.317374.

[15] W. Shafik, "Cyber security perspectives in public spaces: Drone case study," In Festus Fatai Adedoyin and Bryan Christiansen (eds.). *Handbook of Research on Cybersecurity Risk in Contemporary Business Systems*, pp. 79–97, 2023, IGI Global, https://doi.org/10.4018/978-1-6684-7207-1.ch004.

[16] Z. Yang, L. Jianjun, H. Faqiri, W. Shafik, A. Talal Abdulrahman, M. Yusuf and A. M. Sharawy, "Green internet of things and big data application in smart cities development," *Complexity*, 2021, 15 pages, 2021, https://doi.org/10.1155/2021/4922697.

[17] K. Baum, S. Mantel, E. Schmidt and T. Speith, "From responsibility to reason-giving explainable artificial intelligence," *Philosophy and Technology*, 35(1), p. 12, 2022, https://doi.org/10.1007/s13347-022-00510-w.

[18] T. Veiga, H. A. Asad, F. A. Kraemer and K. Bach, "Towards containerized, reuse-oriented AI deployment platforms for cognitive IoT applications," *Future Generation Computer Systems*, 142:4–13, 2023, https://doi.org/10.1016/j.future.2022.12.029.

[19] X. Jin, X. Wang, X. Cao and C. Xue, "Construction and recognition of acoustic ID of ancient coins based on deep learning of artificial intelligence for audio signals," *Heritage Science*, 11(1), p. 46, 2023, https://doi.org/10.1186/s40494-023-00891-x.

[20] Y. G. Butler, "Language education in the era of digital technology," *JALT Journal*, 44(1), p. 137, 2022, https://doi.org/10.37546/JALTJJ44.1-7.

[21] R. Kumar, N. Grover, R. Singh, S. Kathuria, A. Kumar and A. Bansal, "Imperative role of artificial intelligence and big data in finance and banking sector," In *2023 International Conference on Sustainable Computing and Data Communication Systems*, Erode, India, pp. 523–527, 2023, https://doi.org/10.1109/ICSCDS56580.2023.10105062.

[22] S. Knapič, A. Malhi, R. Saluja and K. Främling, "Explainable artificial intelligence for human decision support system in the medical domain," *Machine Learning and Knowledge Extraction*, 3(3), pp. 740–770, 2021, https://doi.org/10.3390/make3030037.

[23] H. Hu, J. Xu, M. Liu and M. K. Lim, "Vaccine supply chain management: An intelligent system utilizing blockchain, IoT and machine learning," *Journal of Business Research*, 156:113480, 2023, https://doi.org/10.1016/j.jbusres.2022.113480.

[24] W. Shafik, S. M. Matinkhah, F. Shokoor and L. Sharif, "A reawakening of machine learning application in unmanned aerial vehicle: Future research motivation," *EAI Endorsed Transactions on Internet of Things*, 8:29, 2022, https://doi.org/10.4108/eetiot.v8i29.987.

[25] Y. Jun, A. Craig, W. Shafik and L. Sharif, "Artificial intelligence application in cybersecurity and cyberdefense," *Wireless Communications and Mobile Computing*, 2021. 10 pages, 2021, https://doi.org/10.1155/2021/3329581.

[26] K. D. Gronwald, "Machine learning, deep learning und artificial intelligence," In Klaus-Dieter Gronwald (ed.). *Globale Kommunikation und Kollaboration: Globale Supply Chain Netzwerk-Integration, interkulturelle Kompetenzen, Arbeit und Kommunikation in virtuellen Teams*, pp. 85–107, 2023, Wiesbaden: Springer Fachmedien Wiesbaden. https://doi.org/10.1007/978-3-658-39099-0_6.

[27] Y. A. Qadri, A. Nauman, Y. Zikria, Y. Bin, A. V. Vasilakos and S. W. Kim, "The future of healthcare internet of things: A survey of emerging technologies," *IEEE Communications Surveys and Tutorials*, 22(2):1121–1167, 2020, https://doi.org/10.1109/COMST.2020.2973314.

[28] L. Zhao, D. Zhu, W. Shafik, S. M. Matinkhah, Z. Ahmad, et al., "Artificial intelligence analysis in cyber domain: A review," *International Journal of Distributed Sensor Networks*, 18(4), 2022, https://doi.org/10.1177/15501329221084882.

[29] Y. B. Bakare, "Machine learning-based smart irrigation system and soil nutrients analysis to increase productivity in agriculture field," *AIP Conference Proceedings*, 2523, 2023, p. 020030. https://doi.org/10.1063/5.0116007.

[30] M. M. Kamruzzaman, S. Alanazi, M. Alruwaili, N. Alshammari, S. Elaiwat, et al., "AI-and IoT-assisted sustainable education systems during pandemics, such as covid-19, for smart cities," *Sustainability*, 15(10):8354, 2023, https://doi.org/10.3390/su15108354.

[31] K. Kumar, P. Kumar, D. Deb, M. L. Unguresan and V. Muresan, "Artificial intelligence and machine learning based intervention in medical infrastructure: A review and future trends," *Healthcare*, 11(2):207, 2023, https://doi.org/10.3390/healthcare11020207.

[32] H. Yu, S. Luo, J. Ji, Z. Wang, W. Zhi, N. Mo, et al., "A deep-learning-based artificial intelligence system for the pathology diagnosis of uterine smooth muscle tumor," *Life*, 13(1):3, 2022, https://doi.org/10.3390/life13010003.

[33] Q. Wu, H. Ma, J. Sun, C. Liu, J. Fang, et al., "Application of deep-learning-based artificial intelligence in acetabular index measurement," *Frontiers in Pediatrics*, 10, p. 1049575, 2023 . https://doi.org/10.3389/fped.2022.1049575.

[34] S. Aminizadeh, A. Heidari, S. Toumaj, M. Darbandi, N. J. Navimipour, et al., "The applications of machine learning techniques in medical data processing based on distributed computing and the Internet of Things," *Computer Methods and Programs in Biomedicine*, 107745, 241, 2023, https://doi.org/10.1016/j.cmpb.2023.107745.

[35] W. Shafik, S. M. Matinkhah, F. Shokoor, and M. N. Sanda, "Internet of things-based energy efficiency optimization model in fog smart cities," *International Journal on Informatics Visualization*, 5(2), pp. 105–112, 2021, https://doi.org/10.30630/joiv.5.2.373.

[36] M. Ryo, "Explainable artificial intelligence and interpretable machine learning for agricultural data analysis," *Artificial Intelligence in Agriculture*, 6, pp. 257–265, 2022. https://doi.org/10.1016/j.aiia.2022.11.003.

[37] W. Shafik, S. M. Matinkhah and M. Ghasemzadeh, "Theoretical understanding of deep learning in uav biomedical engineering technologies analysis," *SN Computer Science*, 1(6), pp. 1–13, 2020, https://doi.org/10.1007/s42979-020-00323-8.

[38] W. Shafik, "Wearable medical electronics in artificial intelligence of medical things," In Agbotiname Lucky Imoize, Valentina Emilia Balas, Vijender Kumar Solanki, Cheng-Chi Lee, Mohammad S. Obaidat (eds.). *Handbook of Security and Privacy of AI-Enabled Healthcare Systems and Internet of Medical Things*, CRC Press, Boca Raton, pp. 21–40, 2024, https://doi.org/10.1201/9781003370321-2.

[39] W. Shafik, S. M. Matinkhah and F. Shokoor, "Cybersecurity in unmanned aerial vehicles: A review," *International Journal on Smart Sensing and Intelligent Systems*, 16(1), pp. 1–16, 2023, https://doi.org/10.2478/ijssis-2023-0012.

[40] S. S. Alghamdi, "The application of artificial intelligence in detecting breast lesions with medical imaging: A literature review," *International Journal of Biomedicine*, 13(1), pp, 9–13, 2023, https://doi.org/10.21103/Article13(1)_RA1.

[41] H. Hua, Y. Li, T. Wang, N. Dong, W. Li, et al., "Edge computing with artificial intelligence: A machine learning perspective," *ACM Computing Surveys*, 55(9):1–35, 2023, https://doi.org/10.1145/3555802.

[42] M. Soori, B. Arezoo and R. Dastres, "Artificial intelligence, machine learning and deep learning in advanced robotics, A review," *Cognitive Robotics*, 3:54–70, 2023, https://doi.org/10.1016/j.cogr.2023.04.001.

[43] L. Rubinger, A. Gazendam, S. Ekhtiari and M. Bhandari, "Machine learning and artificial intelligence in research and healthcare," *Injury*, 54:S69–S73, 2023, https://doi.org/10.1016/j.injury.2022.01.046.

[44] A. Gaurav, B. B. Gupta and P. K. Panigrahi. "A comprehensive survey on machine learning approaches for malware detection in IoT-based enterprise information system," *Enterprise Information Systems*, 17(3):2023764, 2023, https://doi.org/10.1080/17517575.2021.2023764.

[45] N. Prazeres, R. L. Costa, L. Santos and C. Rabadão, "Engineering the application of machine learning in an IDS based on IoT traffic flow," *Intelligent Systems with Applications*, 17:200189, 2023, https://doi.org/10.1016/j.iswa.2023.200189.

III

IoT-Based Techniques for Smart Future Architectures

CHAPTER 8

Internet of Metaverse Things (IoMT)

Applications, Technology Challenges and Security Consideration

Muhammad Shahid Anwar

Gachon University

Wadee Alhalabi

King Abdulaziz University

Ahyoung Choi

Gachon University

Inam Ullah

Gachon University

Ahad Alhudali

King Abdulaziz University

8.1 INTRODUCTION TO THE METAVERSE AND IoT

This section will discuss and address the convergence of the Metaverse with the Internet of Things (IoT) briefly. The Metaverse is a networked virtual environment in which users engage and interact, whereas IoT is the interconnection of actual items. Understanding how they interact is critical for exploring future possibilities. As virtual worlds now progressively resemble our real world, the era of the Metaverse has overtaken the one in which owning a computer was regarded as a luxury. Notably, Facebook, now known as Meta

DOI: 10.1201/9781032648309-11

Platforms, has been at the forefront of developing technology for virtual worlds and has become extremely well known and influential in new markets. As a result, global business units, tech industries, and researchers are actively involved in the Metaverse's evolution into a fully-fledged corporate area. Tech specialists have identified a preferred location within the Metaverse for experimentation, looking at fresh approaches to take advantage of its potential for industrial innovation. The IoT has grown four times more significantly in this environment since this growth cannot be achieved in isolation. This chapter discusses briefly the convergence of IoT and Metaverse with significant applications in different sectors. Figure 8.1 shows the complete outline of this chapter.

8.1.1 Defining the Metaverse

The Metaverse is a digital universe that is virtual, interconnected, and immersive, allowing users to interact, create, and engross with a range of artificial environments, Augmented Reality (AR), Virtual Reality (VR), Mixed Reality (MR), Extended Reality (XR), and experiences in real time. It blurs the distinction between the real and digital worlds, opening up new avenues for social engagement and enjoyment. In today's digital world, the term "Metaverse" is the current buzzword. It is a novel internet application and social world that incorporates emerging technologies such as Artificial Intelligence, Blockchain (BC), and others. The Metaverse can be thought of as a virtual world that mimics reality. Several activities that can be completed in the actual world, such as playing games, shopping, going to concerts, meeting new people, and so on, can be done differently in the Metaverse.

The Metaverse is defined differently depending on users' and participants' points of view and intent. The often debated Metaverse, on the other hand, is a virtual environment that is similar to the actual world and environment in that it is a virtual space for communicating with other users in the environment. Snow Crash launched the Metaverse in 1992 [1], and it was widely researched as the second Life environment in 2006 [2]. Several Metaverse-based applications such as ZEPETO and Roblox have recently received numerous attention. There are four key distinctions between the present Metaverse and the prior second-life Metaverse: (1) The novel Metaverse is more realistic and immersive than the old one and has strong recognition accuracy and performance and a natural cohort model thanks to the advancement of deep learning and AI; (2) Contrasting the former Metaverse, i.e., PC-based, the present Metaverse makes use of mobile devices to improve continuity and accessibility; and (3) With the advancement of security considerations such as virtual currency (e.g., Bitcoin, Dime) and blockchain, the economic proficiency and firmness of Metaverse services have enhanced; and (4) As a result of the restrictions of offline social action such as COVID-19, the interest in virtual reality and the world has advanced.

8.1.2 Understanding the Internet of Things

The IoT is a novel paradigm that lets electronic gadgets and sensors interconnect via the Internet to advance our lives. IoT technology uses intelligent devices and the internet to bring creative solutions to many public/private enterprises, businesses, and governmental organizations around the world [3]. IoT is progressively becoming a significant component of our lives, and it can be felt all around us. The IoT is a game-changing idea that connects

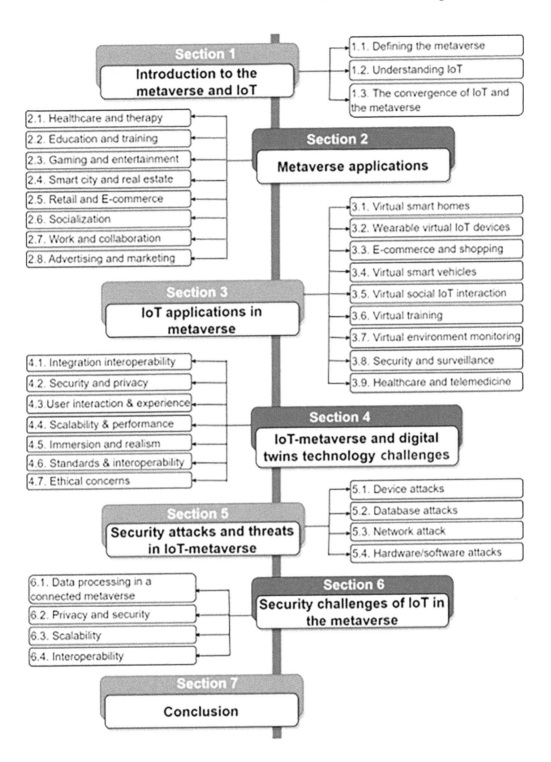

FIGURE 8.1 Chapter outline.

common objects to the internet, allowing for data sharing and automation. It enables gadgets to collect, share, and act on data, resulting in a smarter, more efficient environment. Understanding the IoT is critical for capitalizing on its potential across industries, improving lives, and increasing productivity. IoT has a multidisciplinary vision to help several areas such as transportation, public/private, the environment, industry, medical, and smart grid [4] to name a few. Researchers have discussed the IoT in different ways based on their specific interests and concerns [5–8].

The popular area of IoT application is smart cities, which also includes smart houses that are made up of IoT-enabled daily life home appliances such as air-conditioning, television, video/audio streaming devices, heating system, and security arrangements [9]. These devices communicate with one another to deliver optimal comfort, security, and energy usage. All of this communication takes place over the Internet via an IoT-based central control unit. Some potential IoT applications for smart cities are traffic management, public safety solutions, air quality control, smart parking, smart lightening, and smart waste collection.

8.1.3 The Convergence of IoT and the Metaverse

In recent years, the IoT and Metaverse have been explored and investigated to provide users with more immersive and realistic cyber-virtual experiences in MR environments. As we define it, the IoT is a resilient digital network of interconnected physical items that we use every day. Make use of the link between the physical and technical worlds. In IoT, billions of technical gadgets exchange information, communicate, and access data resources via wired or wireless channels. IoT is a critical component of the Metaverse infrastructure, relying on it to reach its full potential. The integration of IoT and the Metaverse could create new prospects for technological growth and development. IoT and Metaverse are technologies that are concerned with transforming the way we look, interact, communicate, connect, and do our daily tasks. Both technologies require internet access to function and modify the underlying structure of human relationships via IoT-based gadgets. It provides users with a fresh immersive experience for accomplishing the same activities they've been doing on traditional platforms. Figure 8.2 presents the future IoMT scenario.

The IoT can be used in the Metaverse [10], which maps real-time IoT data from real life into a digital reality in the virtual world, to facilitate wireless and seamlessly connected immersive digital experiences. The IoT can enhance users' sensory interfaces in the virtual environment provided by AR and VR [11]. Considering the healthcare application of IoT, the medical health devices can be mounted to the patient's body or a sensor-laden body suit to instrument the patient's state, such as health issues that may evoke a virtual response [12].

Integrating the Metaverse and the Internet of Things brings up new opportunities for industrial sectors, individual needs, and societal demands. The scope of this convergence will support Metaverse in overcoming its limits and growing its applicability into more diverse domains. IoT allows virtual environments to interact and communicate with the real world in real time. In contrast, Metaverse technology delivers the 3D user interface required for IoT device clusters. This provides consumers with an IoT and Metaverse

FIGURE 8.2   Internet of Metaverse Things (IoMT) scenario.

experience that is centered on them. This combination allows for optimum data flow to facilitate data-driven decision-making while requiring less training and effort. In other words, IoT will be the essential link enabling individuals to move seamlessly between the Metaverse and the real world. Furthermore, for a better user experience in the Metaverse, it is necessary to develop a far-advanced IoT technology that can easily support virtual space activities and complexities.

With the integration of IoT and Metaverse, the e-commerce virtual experiences of the virtual fitting room will also be improved, where IoT devices are utilized to detect the user's body movement. Data from images obtained on the user's mobile phone or other smart weighing scales, for example, might be used to update the user's personal body information. This allows Metaverse users to fully immerse themselves in a virtual depiction of the store, overcoming the experience limitations associated with traditional online purchasing. Furthermore, IoT data can be used by the latest Tactile Internet, which creates a network or network of networks for humans or machines to remotely access or control real or virtual things in real time [13,14]. IoT data can give AR/VR applications perspective and contextual awareness of physical objects while also initiating data interchange between the digital and real worlds [15]. For this purpose, AR attached device can respond to the user's finger motions or initiate a cyber-physical function in response to a physical occurrence.

The Metaverse is a linked network of popular 3D simulations and worlds in which the users are denoted by virtual avatars with a sense of social presence and geographical awareness and engage in a huge virtual economy. IoT is critical for connecting the Metaverse to real-world objects or devices. When linked gadgets in the Metaverse can smoothly share and receive information resulting in an extra well-organized copy of the actual environment. IoT sensors connect the physical and virtual worlds, and they play a critical role in data capture from physical assets. Digital twins (DTs) are necessary to virtualize a physical object. To join the virtual world, XR devices are required, and IoT semiconductors are the key component.

In addition, edge computing and 5G are critical enablers of Metaverse technology. Various businesses are progressively financing to spot themselves in the Metaverse market, and the Metaverse's diverse applications can lead to new commercial prospects. The Metaverse is quickly spreading across multiple businesses, and related technology is expected to offer income opportunities across a wide range of verticals. According to the research, the combination of IoT and Metaverse will create new prospects for growth and development in the digital industry, such as IoT companies extending and improving their capabilities and forming strategic collaborations.

The interplay of the real world and the virtual environment enabled by IoT helps to generate a digital twin, a digital mirror of a unique physical thing's physical state and condition [16]. The Metaverse strives to guarantee that the reflection is as near to the real-time physical condition as feasible to establish a viable digital twin. Because of this distinguishing feature, digital twins are becoming one of the most important uses in the Metaverse. In professional situations, digital twins can be built using the Tactile Internet and Haptic Codecs (IEEE P1918.1.1) [17] to establish a group meeting more effectively by allowing users to engage with one another while operating or showing a replica of the hardware or software prototype. Digital twins assist engineers in directly operating 3-D representations of complex systems in technical training programs [18].

This chapter explains how the convergence of IoT and the Metaverse brings the actual world closer to the ever-changing virtual world. This chapter starts with the introduction of IoT and explains the emerging Metaverse technology and potential use cases. It then gives an overview of the Convergence of IoT and the Metaverse. Section 8.2 describes the potential application of Metaverse technologies in different sectors. Section 8.3 presents significant applications that show how IoT can be used in the Metaverse. Furthermore, the IoT-Metaverse and digital twin technology challenges are discussed in Section 8.4. The security attacks and threats in IoT-Metaverse are briefly described in Section 8.5, while the security challenges of IoT-Metaverse are presented in Section 8.6. Finally, Section 8.7 concludes this chapter with some final thoughts and discussion.

8.2 METAVERSE APPLICATIONS

Metaverse intends to set up a cohesive network of three-dimensional virtual and real worlds, in which a single and universal Internet is made available to give users an immersive cyber-virtual experience in physical worlds. Many areas of our daily life could be transformed by the Metaverse. It has the potential to be used for healthcare, gaming, business, education, entertainment, social interactions, and other purposes. We may be able to use the Metaverse in the future to play, work, socialize, and learn, in novel and immersive ways. Two popular applications, AR and VR, in particular, are being developed to provide integrated immersive digital experiences and social connections to Metaverse users. According to Consumer Technology Association projections, the AR and VR industry in medical healthcare is expected to expand from $960 million in 2019 to $7 billion in 2026 as a result of the COVID-19 pandemic [19]. According to Market Study Future's thorough

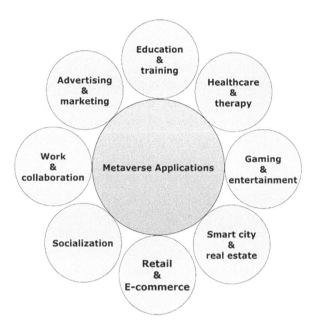

FIGURE 8.3   Applications of Metaverse in different sectors.

study report, the immersive learning or AR/VR-based education market can increase at an 18.2% compound annual growth rate (CAGR) during the next five years (2024 - 2028) [20].

Figure 8.3 depicts the Metaverse's application scenarios, which include healthcare and therapy [21], education and training [22], smart cities [23], gaming and entertainment [24–30], real estate [31], and sociability [32], retail and virtual commerce, virtual collaboration and work, and virtual advertising and marketing.

8.2.1 Healthcare and Therapy

The Metaverse has the potential to play a key role in healthcare and therapy. Medical training can benefit from virtual simulations and environments, which allow healthcare workers to practice procedures and scenarios in a safe and regulated environment. Furthermore, virtual therapy sessions can make mental health assistance and interventions available to people all over the world. According to the World Economic Forum [21], digital services will be one of the most important forces in revolutionizing healthcare over the next decade. Before the COVID-19 pandemic, 43% of healthcare facilities could offer telemedicine, a figure that would rise to 95% by 2020 [33]. The Metaverse has the potential to speed up the growth of telemedicine, which will help patients, doctors, healthcare workers, and medical students. The patients in the Metaverse environment can immerse themselves via their avatar to confer with their doctors [34–38]. Many technological companies such as Microsoft Hololens [39] are creating clinical assisting technologies for on-demand surgical procedures. Furthermore, the Metaverse can aid in psychotherapy [32]. The Metaverse can be used to create a virtual and calming environment in which persons with mental illnesses can communicate and interact with different avatars.

8.2.2 Education and Training

The Metaverse and education sectors are closely connected in that they are always digitally facilitated in modern education. In virtual education, the students share virtual spaces, content, and methods that benefit both their personal development and academic skills. However, many people are suspicious about the Metaverse's significance in education, claiming that it can be misleading in terms of improving the educational process [40]. Metaverse technology has gained popularity in the education and training sector [41–43]. AR/VR will be widely used to change traditional teaching methodologies that provide interactive and immersive learning experiences for students [22]. Traditional teaching methodologies are being transformed by AR/VR that emphasizes visualization-based learning ideas [44]. The Metaverse is projected to expand various educational institutions and organizations that can depict teaching and learning information and increase understanding of learning content to create engaging and immersive learning environments for students. The Metaverse can also be used to build safe but immersive virtual lab conditions. Furthermore, some learning materials, such as the human body's structure and functioning of an organ system, or the universe, are difficult to witness directly or explain in text [45]. AR/VR can be built to effectively provide students with the required constant practice and experience to improve the understandability of the learning materials.

Furthermore, another study [46] argues that the Metaverse contributes to the improvement of the educational process, particularly the teacher-student connection, which, thanks to the Metaverse, is free of time and location constraints. Furthermore, the Metaverse is critical to transforming the traditional static educational model into a dynamic model by mobilizing a diverse range of situations, tools, forms of learning methods, and assessments and placing the student at the center of the educational process, potentially increasing their motivation to learn [46].

8.2.3 Gaming and Entertainment

The potential of Metaverse has been widely recognized in the game industry and entertainment [25–30]. Games in the Metaverse provide players with extensive virtual environments to collaborate with other users and players and go on fascinating adventures. Moreover, the Metaverse surpasses gaming by serving as a platform for immersive entertainment experiences such as 360 movie screenings, virtual concerts, and art exhibitions. Furthermore, the advancement of Metaverse-related technology has significantly boosted gaming immersion, which can dramatically improve user experience, enjoyment, immersion, playability, and usability. Roblox is a Metaverse game that relies primarily on VR technology having a monthly 150 million active users [47]. Besides the extensive use of money and user individual information in the Metaverse, a blockchain-based game has been presented in [48]. The Metaverse can also be used for huge events such as exhibitions, concerts, and book signings because it provides an immersive setting. Recently, Korean pop musicians Asepa and Black Pink published a few new songs and had fan signings in the virtual environment of Metaverse [24].

8.2.4 Smart City and Real Estate

The IoT-based digital twins can leverage the data to digitize real-world objects such as roads and streets, residences, vehicles, and city infrastructures, and develop virtual cities as a supporting technology in the Metaverse. The Metaverse is primarily employed in smart cities to enhance the allocation of services and resources. Real estate has been identified as the Metaverse's next promising application entry [31]. AR/VR, as one of the core technologies powering the Metaverse, offers users a realistic and immersive experience [49]. Furthermore, because it can create an immersive virtual environment and enable real-time interaction, the Metaverse has a high potential for deployment in interior design and architectural areas with the use of AR/VR and AI technologies [50].

8.2.5 Retail and E-Commerce

The Metaverse unlocks new opportunities for retail and e-commerce. Users can browse and purchase digital and actual things, try on virtual outfits, and design their virtual living environments in Metaverse virtual storefronts. Brands may use the Metaverse to create one-of-a-kind and interactive retail experiences that increase customer engagement and happiness. Retailers may use Metaverse to build the most engaging and immersive shopping environments where customers can test and purchase things in real time. Customers will thus have the identical purchasing experience as if they were in a real-world retail mall. To summarize, customers can communicate with shops virtually from anywhere.

According to a recent survey [51], people are particularly interested in the Metaverse's retail sector (i.e., shopping). As a result, to attract digital customers, firms are experimenting with new concepts and media in both real and virtual contexts [52]. Immersive VR media, for example, offers the advantage of offering an experience similar to buying in a physical store. [53] highlighted the importance of new technologies, such as VR in improving the consumer experience of online fashion purchasing, and discussed the necessity for more impressive and better retail experiences that satisfy fashion consumers.

8.2.6 Socialization

The Metaverse surpasses time and space, allowing for numerous forms of social interaction and bringing people closer together. People can seek out remote social networks to foster a more authentic social atmosphere. In terms of human social formations, the Metaverse brings up a new application field [32]. The Metaverse transcends time and space, allowing for numerous forms of social interaction and bringing individuals closer together [54]. People can use virtual offices, and virtual dating, and can pursue virtual meetings to meet higher-level demands beyond the physical world.

The effect of the COVID-19 epidemic has highlighted the relevance of teleworking and remote social networking [55]. To build a more authentic social environment, the Metaverse can compensate for the constraints of existing models and increase the functionality of telecommuting and remote social platforms.

8.2.7 Work and Collaboration

The Metaverse can completely transform how we work and collaborate. Within the Metaverse, virtual office spaces allow distant teams to connect, conduct meetings, and work on projects in a dynamic and immersive environment. It removes the barriers of physical distance, promoting worldwide collaboration and creativity. Based on pilot research done in an academic health informatics laboratory, the use of the Metaverse in work and collaboration has demonstrated great promise. The researchers wanted to know how a Metaverse-based virtual workspace may help lab members communicate and collaborate [55].

8.2.8 Advertising and Marketing

The Metaverse expands the possibilities for advertising and marketing tactics. Brands may offer immersive virtual world experiences that allow users to connect with their products and services in novel ways. Personalized marketing and virtual brand activations have become essential components of the Metaverse experience.

The application of Metaverse can introduce virtual objects, AR and VR showrooms, and branded gaming to promote the business and reach new customers. The advertisers can even create 3D marketing experiences that attract consumers' interest by engaging them in novel ways, assisting them in gaining leads and retaining customers.

8.3 IoT APPLICATIONS IN METAVERSE

The platforms of the Metaverse cannot be operated alone on a personal machine. The Metaverse is essentially an internet-based technology. An improved internet and supported mechanism version would improve its features, while IoT would handle the main requirements of Metaverse platforms. By combining these technologies, people can interact in an altogether new spatial realm where they can enjoy immersive experiences from the comfort of their own homes. Both are required for the Metaverse platforms to function. As a result, they must work together to create an industrial solution.

The IoT can be used in the Metaverse to offer users a more immersive and engaging experience. IoT devices can transfer data from the actual world into the Metaverse, making the experience more authentic. The user's mobility as well as voice commands or gestures might be tracked in the Metaverse with the integration of IoT sensors. Similarly, this integration will also potentially provide haptic feedback along with tracking and authentication of users and prevent unauthorized access to the Metaverse. It can boost and improve the user's experience of the Metaverse. Moreover, it will assist the users in experiencing a more engaging and immersive environment while bridging the actual and Metaverse worlds with improved safety. Significant and notable applications of IoT in the Metaverse are shown in Figure 8.4.

8.3.1 Virtual Smart Homes

Within the Metaverse, users can interact with IoT-enabled virtual smart homes. The user can control home lighting, virtual, appliances, room temperature, and other daily life equipment. Virtual smart houses are simulated versions of real-world dwellings and are

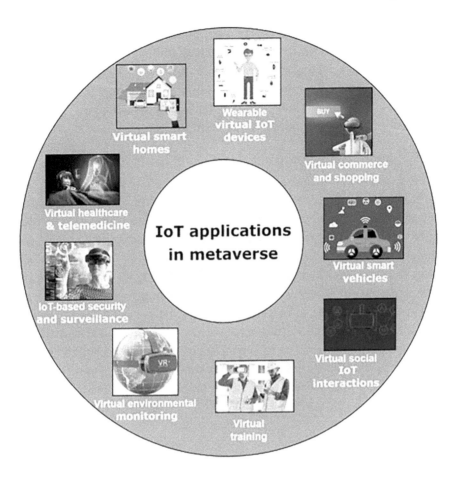

FIGURE 8.4 Applications of IoT in Metaverse.

made possible by IoT devices. These gadgets enable users to communicate with their home appliances in the Metaverse in several ways. These communications can regulate appliances, lighting, temperature, and other home comforts. This can be done by using their VR headset to adjust the thermostat, turn on the lights, or start the coffee maker. The users can also use their voice to control these appliances or engage with them through gestures. Virtual smart houses have numerous conceivable advantages, including the ability to operate them from anywhere in the globe with an internet connection. Users may keep an eye on their houses for security issues like unauthorized entry or equipment problems. Users' virtual homes can be customized to their satisfaction, and they can even create many versions of their homes for different purposes. Virtual smart houses are likely to become increasingly common as the Metaverse evolves. They provide a novel way to engage with our houses, as well as a variety of conveniences and security benefits.

8.3.2 Wearable Virtual IoT Devices

Wearable virtual IoT devices such as smartwatches or AR glasses, might provide users with real-time information or interactions within the Metaverse. Some of the possibilities include vital sign monitoring, gesture control, and immersive haptic feedback. Within the

Metaverse, virtual smartwatches, fitness trackers, and other IoT gadgets can offer users virtual health and activity monitoring. With the integration of IoT and Metaverse, several smart product interactions, virtual representation of IoT devices, real-time data visualization, virtual shopping with IoT products, virtual shopping analytics, IoT-driven events and experiences, enhanced VR and AR experiences, remote virtual shopping, secure transactions, and authentication became possible for the users.

8.3.3 Virtual Commerce and Shopping

With the integration of Metaverse and IoT, virtual commerce and shopping in the Metaverse can provide customers with ever more realistic and interactive shopping experiences. The combination of IoT and the Metaverse enables seamless interactions between the virtual and physical worlds, allowing real-world products and equipment to be integrated into the virtual purchasing experience.

IoT devices could link to Metaverse-based virtual shopping experiences, allowing consumers to digitally try on clothes, accessories, or other things while obtaining real-time product availability and pricing information. It will also allow users to make virtual purchases and transactions through virtual stores with smart shelves and interactive product displays. Despite the full potential of IoT-enabled virtual commerce and shopping in the Metaverse has yet to be realized, the integration of these technologies can pave the way for unique and engaging retail experiences that bring the virtual and physical worlds together in fascinating new ways.

8.3.4 Virtual Smart Vehicles

The integration of IoT technology into virtual representations of automobiles within the Metaverse is referred to as the Metaverse. These virtual smart vehicles can provide consumers with a variety of engaging and immersive experiences, blurring the lines between the real and virtual worlds. Some characteristics and applications of virtual smart automobiles in the IoT Metaverse are virtual test drives, smart vehicle integration with smart homes, IoT-enabled safety features, virtual car shows and events, virtual racing and competitions, remote vehicle monitoring, smart traffic simulation, and virtual car ownership.

The combination of IoT technology and virtual smart automobiles in the Metaverse not only delivers entertainment and gaming experiences, but also allows car manufacturers, software developers, and consumers to explore the potential of connected vehicles in new and inventive ways. It's crucial to remember that the creation of virtual smart automobiles in the IoT Metaverse is a constantly growing topic, and improvements in technology and the Metaverse ecosystem will almost certainly lead to even more intriguing applications in the future.

8.3.5 Virtual Social IoT Interaction

With IoT-enabled Metaverse, virtual social IoT interaction is anticipated to play a larger role in defining how people socialize, engage, and experience both the virtual and physical worlds. Users in the Metaverse can access shared virtual worlds and communicate with one another as avatars. IoT data can be integrated into these virtual environments to provide dynamic and context-aware interactions. With the use of IoT data, it can improve the experience of virtual meetups, virtual social events, and virtual conferences. Besides, in the virtual fitness classes, the participants may have their connected fitness trackers' heart rates and activity levels shown within the Metaverse.

However, privacy and security concerns must be considered to maintain a safe and interesting virtual social experience. These virtual social environments are shared virtual spaces, IoT-enabled avatars, virtual social events, personalized social interactions, and remote social presence.

8.3.6 Virtual Training

Organizations may develop dynamic, adaptive, and highly successful learning experiences by incorporating IoT and the Metaverse into virtual training. This can be useful in industries where hands-on practice and real-time data analysis are essential for skill development and performance improvement. With the convergence of IoT and Metaverse technology, virtual training is expected to become an important component of professional growth in a variety of fields. It can also improve the learning experience for trainees and trainers by merging real-world data and interactions into immersive virtual worlds. This method gives trainees hands-on, engaging, and realistic learning experiences. Such pieces of virtual training can be delivered in many ways including realistic simulations, performance monitoring, IoT-enabled virtual equipment, team training and collaboration, hazard simulation and safety training, and continuous learning and updates. In practice, IoT sensors and equipment can collect real-time data, which is then combined with virtual training simulations. Therefore, trainees may engage with virtual items that respond accurately to IoT device data. Thus, virtual training can replicate dangerous scenarios and safety protocols using IoT data from real-world safety sensors. This allows trainees to practice safe practices in a risk-free virtual environment.

8.3.7 Virtual Environment Monitoring

With the integration of Metaverse and IoT-enabled weather monitoring, virtual surroundings become more dynamic and responsive, offering users a more immersive and engaging experience. This integration can pave the way for new virtual domain applications such as gaming experiences, virtual tourism, environmental education, and training simulations. The virtual world can imitate real-world weather scenarios by using real-time IoT weather data. Users may face different weather conditions and occurrences that correspond to real-world weather, such as changing weather patterns, rain, snow, and wind. Virtual environments can also be built with virtual IoT sensors that function similarly to real-world sensors. air quality, humidity, temperature, and other environmental parameters can be measured by these sensors to produce a more authentic virtual experience. Furthermore, real-time weather integration can improve virtual tourism and travel experiences. Users exploring virtual environments may come upon virtual weather patterns that are similar to the current weather conditions in real-world locales.

8.3.8 Security and Surveillance

IoT and Metaverse integration in security and surveillance applications provides a comprehensive and linked way to monitor and respond to security risks. Organizations and individuals can strengthen their security procedures, increase incident response, and acquire a more holistic perspective of their security environment by combining virtual and real-world security data. Several security and surveillance areas can be improved with the IoT-enabled Metaverse such as virtual security scenarios, virtual surveillance, remote surveillance, incident response and collaboration, and integration with alarm systems. Virtual settings in the Metaverse can be constructed to imitate security scenarios such as virtual patrols, access control, and intrusion detection. Users can take part in security training simulations, which put their responses to various security issues to the test. Security personnel can use the Metaverse for training and simulations to practice responding to various security threats and scenarios, improving their readiness in real-world situations. Furthermore, IoT-enabled security devices such as smart cameras, door locks, and motion sensors can be connected with the Metaverse in the real world. Within the virtual environment, users can remotely manage and monitor these devices.

8.3.9 Virtual Healthcare and Telemedicine

Telemedicine and virtual healthcare using IoT-enabled devices Metaverse provides cutting-edge technology to improve medical services, improve patient experiences, and enable remote healthcare delivery. Healthcare providers can use real-time health data from IoT devices to develop immersive and secure virtual healthcare environments.

Several sectors can benefit from IoT-enabled Metaverse to implement virtual healthcare and telemedicine such as virtual consultations, remote patient monitoring, medical simulation and training, virtual medical records, remote surgeries and consultations, and smart healthcare facilities. Through the Metaverse, healthcare practitioners and patients can conduct virtual consultations, and IoT-enabled medical devices, such as remote patient monitoring systems or wearable health trackers, can convey real-time health data to healthcare experts. Besides, healthcare workers can participate in virtual medical simulations and training exercises. IoT data can be used to create realistic scenarios, allowing trainees to digitally diagnose and treat patients. Furthermore, patients can collect health data at home using IoT-enabled medical devices, such as blood pressure, blood glucose levels, or ECG readings. During telemedicine sessions in the Metaverse, these data can be transmitted to healthcare practitioners.

8.4 IoT-METAVERSE AND DIGITAL TWINS TECHNOLOGY CHALLENGES

The IoT-empowered Metaverse enables users to construct virtual environments and experiences beyond their wildest expectations, as well as accurate duplicates of reality that can bring reality to the digital world [56,57]. In the Metaverse, IoT-based digital twins generate immersive experiences by fusing the virtual and physical worlds. Users may easily engage with smart gadgets, products, and environments thanks to these networked replicas. The huge data ecology of the Metaverse boosts digital twin capabilities, allowing for real-time updates, simulations, and personalized experiences. As the Metaverse develops, IoT-based digital twins will play a critical role in influencing the future of augmented and virtual reality applications. Building facilities, operational procedures, human-computer interaction, and social services are examples of IoT-based digital twins that generate a virtual counterpart of physical things or services. Digital twins are commonly used in the Metaverse to give an immersive shopping experience to customers in the real world. Integrating digital imitations with physical goods and services can also help with data analytics, letting businesses use real-world situations for simulations before making costly decisions. Digital twins, which are digital depictions of real-world objects, can coordinate properties, procedures, and operational systems with the actual objects and everyday activities such as analysis of data, visualization, and predictive modeling [58]. Digital twins play an important role in how the real and virtual worlds interact via IoT links [59]. Fluctuations in the actual world are thus rejected in the virtual world. These one-of-a-kind digital twins could be one of the Metaverse's core building elements, replicating the actual world, including its structure and functions, to act as gateways for consumers to experience and enjoy an interactive and immersive virtual environment.

Engineers and service providers can use digital twins to remotely replicate virtual items of technologies and processes and do some physical analysis [60]. Since David Gelernter originally presented the concept of digital twins in his book [61], digital twins have provided businesses with a unique view into how goods are built, run, and performed, making it easier to provide superior products or services. Metaverse employs digital twins to increase productivity and profitability throughout the development of the product life cycle, from design to monitoring, servicing, and post-production. Digital twins can be utilized to develop virtual simulated prototypes for Metaverse and provide vast quantities of predictive data about performance results during the product ideation phase of Metaverse. Make essential product adjustments based on anticipated outcomes to match preset standards before investing in a physical prototype. This saves firms time and money by minimizing the number of iterations needed to bring a Metaverse product to market.

Although IoT-based digital twins and Metaverse are gaining attention, digital twins are useless unless IoT networks are constantly modernized with real-time data, a challenge that ML and AI have been grappling with for quite some time. The Metaverse is a gathering place for digital twins and IoT. The users are looking forward to using AR and VR applications with engaging and immersive experiences in future [62].

Due to the complexity of combining real-world items and systems with virtual settings, developing digital twins within the Metaverse involves various challenges. Among the most common challenges, Figure 8.5 shows some common problems that developers may encounter.

8.4.1 Data Integration and Interoperability

It can be difficult to integrate real-time data from physical objects or systems into digital twins within the Metaverse. Ensuring compatibility and interoperability among various data formats, protocols, and devices can provide technological challenges. It is critical

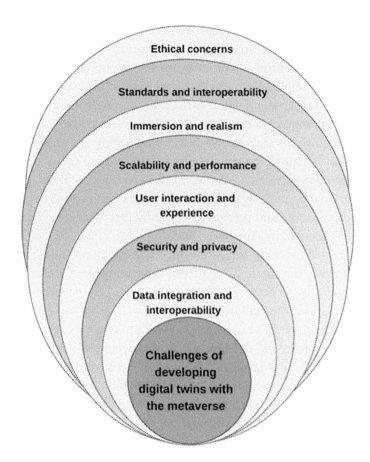

FIGURE 8.5   Developing challenges of digital twin with Metaverse.

to develop robust data pipelines and standardization procedures to ensure seamless data integration.

8.4.2 Security and Privacy

Digital twins sometimes entail sensitive data, such as physical asset information or personal user data. It is vital to ensure the security and privacy of these data throughout the digital twin-Metaverse ecosystem. To defend against unauthorized access, data breaches, and privacy violations, strong security mechanisms such as authentication, encryption, and access controls must be implemented.

8.4.3 User Interaction and Experience

Creating interactive and engaging user experiences within the Metaverse is critical for user adoption. To ensure natural interactions between users and their digital twins, as well as other users, smart interface design, haptic feedback, natural language processing, and gesture detection are required. A significant aspect is balancing usability, functionality, and immersion.

8.4.4 Scalability and Performance

Creating digital twins capable of supporting a high number of concurrent users and complicated interactions within the Metaverse may exhaust system resources. Scaling the infrastructure to handle an increasing user population while retaining ideal performance, such as low latency and excellent responsiveness, can be difficult.

8.4.5 Immersion and Realism

Offering a complete and high level of immersion and realism in the Metaverse can be difficult. Advanced rendering techniques, physics simulations, and thorough object modeling are required to create realistic and detailed virtual worlds that accurately mimic the physical counterparts of digital twins. It is critical for an immersive experience to strive for a seamless integration of the actual and virtual worlds.

8.4.6 Lack of Standards and Interoperability

With numerous technology platforms emerging in the field, a lack of standardization and interoperability might be a problem. A cohesive and connected ecosystem requires compatibility and easy integration of various digital twin platforms, Metaverse environments, and gadgets. The creation of industry-wide standards and processes can aid in addressing this issue.

8.4.7 Ethical Concerns

As digital twins grow more common and integrated with the Metaverse, ethical concerns emerge. Data ownership, data privacy, algorithmic prejudice, and the possible misuse of digital twins are all issues that must be addressed. It is critical to establish ethical norms and frameworks to control the production and use of digital twins in the Metaverse.

8.5 SECURITY ATTACKS AND THREATS IN IoT-METAVERSE

The IoMT faces many challenges and numerous privacy issues that arise when IoT and the Metaverse meet. This chapter aims to underline the need of protecting user privacy and ensuring that the Metaverse environment respects people's rights while providing immersive experiences. Addressing these privacy concerns is critical for fostering confidence and ethical use of IoT-enabled Metaverse technology. Shared user data and 3-D virtual worlds are interconnected with the IoT system in the Metaverse and integrated for remote virtualization. The Metaverse will mostly certainly integrate services of IoT systems, applications, or from various firms, institutions, or organizations [63]. As a result, the biggest security and privacy concerns stem from the fact that the integration of various IoT devices necessitates all allies to coordinate and network data with one another.

There are possible security risks and assaults that could target the various interconnected digital environments and devices that make up the Metaverse. Figure 8.6 shows several possible threats to the Metaverse environment and connected device attacks to consider.

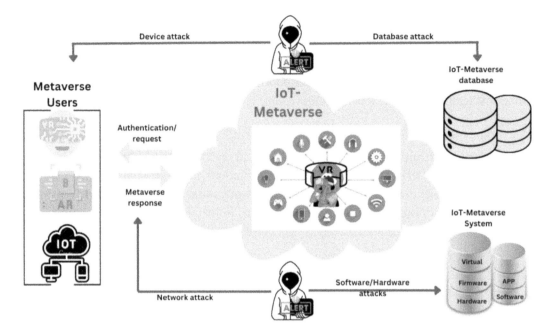

FIGURE 8.6   IoT-Metaverse security and privacy threats.

8.5.1 Device Attack

The hardware and software flaws of devices used to enter the Metaverse are the target of device attacks. These kinds of attacks can compromise a user's device or possibly result in greater exploitation. Malware and ransomware are critical device attacks that infect Metaverse devices with harmful software to steal data, interfere with functioning, or demand a ransom in exchange. Another serious device attack is identity theft and impersonation that pretend to be the real user and stealing credentials for authentication from devices to use them in the Metaverse. Similarly, device bricking is unauthorized firmware updates or alterations that render Metaverse devices inoperable. Another device attack is sensory manipulation that alters the sensory inputs offered by VR/AR systems to cause users uncomfortable, harm, or confuse them.

8.5.2 Database Attacks

Database attacks try to obtain unauthorized access to or manipulate stored information in the Metaverse's data storage systems. These kinds of attacks can result in virtual asset loss, data breaches, and other security difficulties. One of the database attacks is SQL Injection which is a serious attack. This attack can gain unauthorized access to data or execute malicious commands by exploiting weaknesses in the Metaverse's database queries. Another database attack is data theft that steals user profiles, in-game items, virtual cash, and other important data from the Metaverse's databases. Moreover, data manipulation attack manipulates the data in the Metaverse's databases, resulting in discrepancies, disruptions, or illegal changes.

8.5.3 Network Attack

The network attacks target the systems that connect the Metaverse's interconnected gadgets. These kinds of attacks try to stop or disrupt data flow, harming the confidentiality, availability, and integrity of data. Distributed Denial of Service (DDoS) is a network attack that flooded the servers on the Metaverse network with excessive traffic to prevent users from using services. Similarly, the Man-in-the-Middle (MitM) intercepts the user-server communications to eavesdrop, alter, or introduce harmful content. Another network attack in the Metaverse environment could be packet sniffing that could reveal sensitive data capturing packet data transmitted between users. Network eavesdropping is the passive observation of network traffic to obtain data on user interactions and activity.

8.5.4 Hardware and Software Attack

The hardware and software attacks compromise the Metaverse user experience, functionalities, and security. These attacks may take advantage of flaws in the Metaverse's applications or underlying infrastructure. One of the software attacks is Zero-Day Exploits which targets the previously unknown vulnerabilities in Metaverse software before developers can fix them. Backdoor entry is another attack that introduces invisible access points into Metaverse hardware or software to get unauthorized control. Another attack is Supply Chain Attacks that attempt to inject harmful components into the Metaverse ecosystem by compromising the hardware or software supply chain. Furthermore, manipulation of firmware is also the hardware and software attack that changes the firmware of devices or infrastructure components to obtain control or access.

A thorough approach to cybersecurity is required to protect the Metaverse from these and other potential threats. Strong encryption, prompt vulnerability patching, regular security audits, user education, and collaboration among Metaverse device manufacturers, platform developers, and security experts are all part of this. As the Metaverse evolves, remaining diligent and proactive in dealing with security concerns will be critical to providing a safe and entertaining virtual environment.

8.6 SECURITY CHALLENGES OF IoT-METAVERSE

Although the study of IoT-based Metaverse applications and technologies can be used to empower the Metaverse with IoT, continued research and implementation of IoT-empowered Metaverse must address the following important open concerns. VR and AR are the two most popular Metaverse technologies [64,65]; however, they offer security threats, particularly to user privacy. To tackle these threats, clarifications to questions such as how hacked AR devices influence privacy, how AR manufacturers utilize and safeguard user data, and if its sharing is required with other parties [64]. Each of these problems raises questions concerning Metaverse security, such as denial of service, social engineering attacks, and credential theft [66,67]. While virtual reality delivers immersive experiences, it also cuts off people from the real world. When consumers are disconnected from reality, they are vulnerable to physical security threats. Polarization and radicalization are two more significant security issues related to the Metaverse [64]. Harassing, Cyberbullying, and trolling are privacy and security problems that arise from the Metaverse's radical behavior and polarization [64]. To enable the widespread adoption of the Metaverse, significant attention must be paid to all of the aforementioned security and privacy concerns.

8.6.1 Data Processing in a Connected Metaverse

The Metaverse, like the Internet, is vulnerable to a variety of security concerns. It is hard to discuss the Metaverse's future without addressing cybersecurity concerns. Although Internet and Metaverse risks are quite similar, dealing with threats in a virtual environment can be incredibly difficult and costly. Businesses and individuals in the Metaverse confront numerous security challenges. Metaverse applications require real-time IoT data collection to study the activities or states of real-world items or actions [68]. It is quite difficult to monitor items and systems and collect data from all facets, together with operational information, monitoring information, initial data, business data, and program data [58]. According to [69], data analysis and data mining must be performed after collecting real-time IoT data. Contrasting the traditional simulations [70,71], the Metaverse may run numerous simulation procedures with real-time data and provide real-time responses to the sender object for improvement. To provide users/engineers with a complete picture of how the virtual world is working, these simulation procedures depend on data modeling methods such as engineering simulations [72,73], physical analysis [74,75], machine learning [76], neuromorphic computing [77], and data mining [74]. Furthermore, most consumer AR/VR systems contain a variety of sensors that create a massive quantity of data in various modalities and categories. It is worth noting that the Metaverse can use data fusion methods from several data sources to generate a complete view (e.g., automobile driving and road conditions), rather than multiple independent perspectives [58,78]. Storage, transport, security, and privacy of dense information are key data management challenges [79,80].

8.6.2 Scalability

The difficulties and obstacles that develop while attempting to handle increasing demands and expansion in the interconnected virtual environment are referred to as scalability challenges and issues in an IoT-based Metaverse. As the Metaverse grows and accommodates more IoT devices, users, and interactions, numerous scaling issues may arise. Some of the major obstacles and issues are network congestion, data processing and storage, latency and response time, device heterogeneity, scalable infrastructure, resource management, content delivery, and user experience. The popularity of Metaverse has been growing and named one of the future's top five evolving technologies in the coming 10 years. The investment in IoT development is predicted to increase from $12 billion in 2020 to $72.8 billion in 2024 globally. Over 200 big firms including Samsung, JP Morgan, and Nike have already applied and moved some of their customer-facing activities to the Metaverse. It is possible that an increasing number of IoT devices would be able to develop and populate virtual worlds for users in social media with reasonable ease and minimal barriers to service access. Scalability will be a big challenge with any Metaverse platform as IoT applications expand.

To increase scalability, the Metaverse must be built on a decentralized architecture. In real-time interaction, the user's location and scene status must be saved offline. To address these scaling obstacles and issues, careful design, optimization, and the utilization of cutting-edge technology are required. Creating a sustainable and immersive virtual environment capable of supporting the demands of a growing user base and diverse IoT ecosystem can be aided by designing the IoT-based Metaverse with scalability in mind from the start.

8.6.3 Interoperability

When integrating IoT devices into the Metaverse, interoperability and consistency of virtual platforms are critical difficulties. These difficulties include enabling flawless communication and interaction between various IoT devices and virtual environments, as well as preserving consistency across multiple platforms. One of the Metaverse's significant accomplishments is the creation of a virtual environment [81] where the participants can join and collaborate on various activities such as gaming, viewing movies, and other work. According to the current scenario, more than one company is creating Metaverse platforms [82]. Facebook, Microsoft, and Apple are prominent examples [83]. Different devices are also required for IoT networks to connect to many Metaverse platforms and communicate with various groups. Furthermore, establishing interoperability [84] comprises compatibility across diverse Metaverse events, places, and measures, as well as interoperability of mixed networks in the platform.

Interoperability includes guaranteeing user and platform compatibility in Metaverse along with the compatibility between platforms and operating systems. There are several challenges and obstacles such as cross-platform content sharing, device variety, protocol integration, virtual environment, data format standardization, virtual environment, and synchronization and timing. Resolving these difficulties requires a combination of technical solutions, industry collaboration, and the formation of standards and conventions. Implementing open standards and supporting interoperable design principles would aid in the formation of a more united and coherent IoT-based Metaverse, boosting user experiences and stimulating innovation.

8.7 CONCLUSION

This chapter presents the Internet of Metaverse Things and its groundbreaking vision of the future, where the boundaries between the virtual and real-world blur, and where the integration of IoT and the Metaverse brings forth an unparalleled level of immersion, intelligence, and interconnectedness. By shedding light on the potential applications of IoT in Metaverse. Moreover, this chapter discusses and explains the Metaverse applications that will bring major changes in the industry and address the associated challenges. The transformation into IoMT provides users with unparalleled possibilities, setting the path for disruptive applications in a variety of industries such as healthcare, education, gaming, smart cities, and others. While IoMT offers doors to new experiences, obstacles such as technological barriers, security concerns, and regulatory requirements loom. As IoMT reshapes our digital ecosystem, addressing challenges such as privacy, data security, scalability, and interoperability will be critical. Embracing this transition has enormous promise, but it will require joint efforts to create a secure and prosperous IoMT ecosystem.

REFERENCES

[1] N. Stephenson, *Snow Crash*. Bantam Books, New York, USA, 1992.

[2] S.-M. Park and Y.-G. Kim, "A metaverse: Taxonomy, components, applications, and open challenges." *IEEE Access*, vol. 10, pp. 4209–4251, 2022.

[3] A.R. Sfar, C. Zied, and Y. Challal, "A systematic and cognitive vision for IoT security: A case study of military live simulation and security challenges." In: Proceedings of the 2017 International Conference on Smart, Monitored and Controlled Cities (SM2C), Sfax, Tunisia, 17–19 February, 2017. https://doi.org/10.1109/sm2c.2017.8071828.

[4] W. U. Khan, N. Imtiaz, and I. Ullah, "Joint optimization of NOMA-enabled backscatter communications for beyond 5G IoT networks." *Internet Technology Letters*, vol. 4, no. 2, p. e265, 2021.

[5] M. Asif, W. U. Khan, H. M. Rehan Afzal, J. Nebhen, I. Ullah, A. Ur Rehman, and M. K. A. Kaabar, "Reduced-complexity LDPC decoding for next-generation IoT networks." *Wireless Communications and Mobile Computing*, vol. 2021, pp. 1–10, 2021.

[6] D. Gupta, S. Juneja, A. Nauman, Y. Hamid, I. Ullah, T. Kim, E. M. Tag eldin, and N. A. Ghamry, "Energy saving implementation in hydraulic press using industrial Internet of Things (IIoT)." *Electronics*, vol. 11, no. 23, p. 4061, 2022.

[7] T. Mazhar, H. M. Irfan, I. Haq, I. Ullah, M. Ashraf, T. Al Shloul, Y. Y. Ghadi, Imran, and D. H. Elkamchouchi, "Analysis of challenges and solutions of IoT in smart grids using AI and machine learning techniques: A review." *Electronics*, vol. 12, no. 1, p. 242, 2023.

[8] H. U. Khan, M. Sohail, F. Ali, S. Nazir, Y. Y. Ghadi, and I. Ullah, "Prioritizing the multi-criterial features based on comparative approaches for enhancing security of IoT devices." *Physical Communication*, vol. 59, p. 102084, 2023.

[9] T. Mazhar, D. B. Talpur, T. Al Shloul, Y. Y. Ghadi, I. Haq, I. Ullah, K. Ouahada, and H. Hamam, "Analysis of IoT security challenges and its solutions using artificial intelligence." *Brain Sciences*, vol. 13, no. 4, p. 683, 2023.

[10] T. G. Kanter, "The metaverse and extended reality with distributed IoT." IEEE IoT Newsletter, November 2021. [Online]. Available: https://iot.ieee.org/newsletter/november-2021/the-Metaverse-andextended-reality-with-distributed-iot.

[11] N. Pereira, A. Rowe, M. W. Farb, I. Liang, E. Lu, and E. Riebling, "Arena: The augmented reality edge networking architecture." In: *Proceedings of the IEEE International Symposium on Mixed Augmented Reality (ISMAR)*, Bari, Italy, 2021, pp. 479–488.

[12] A. H. Sodhro, S. Pirbhulal, and A. K. Sangaiah, "Convergence of IoT and product life-cycle management in medical health care." *Future Generation Computer Systems*, vol. 86, pp. 380–391, 2018.

[13] N. Promwongsa et al., "A comprehensive survey of the tactile Internet: State-of-the-art and research directions." *IEEE Communications Surveys and Tutorials*, vol. 23, no. 1, pp. 472–523, 2020.

[14] A. Aijaz and M. Sooriyabandara, "The tactile Internet for industries: A review." *Proceedings of the IEEE*, vol. 107, no. 2, pp. 414–435, 2019.

[15] E. Lu, J. Miller, N. Pereira, and A. Rowe, "FLASH: Video-embeddable AR anchors for live events." In: *Proceedings of the IEEE International Symposium on Mixed Augmented Reality (ISMAR)*, Bari, Italy, 2021, pp. 489–497.

[16] R. Minerva, G. M. Lee, and N. Crespi, "Digital twin in the IoT context: A survey on technical features, scenarios, and architectural models." *Proceedings of the IEEE*, vol. 108, no. 10, pp. 1785–1824, 2020.

[17] E. Steinbach et al., "Haptic codecs for the tactile Internet." *Proceedings of the IEEE*, vol. 107, no. 2, pp. 447–470, 2019.

[18] N. Stojanovic and D. Milenovic, "Data-driven digital twin approach for process optimization: An industry use case." In: *Proceedings of the IEEE International Conference on Big Data (Big Data)*, Seattle, WA, USA, 2018, pp. 4202–4211.

[19] M. Slovick. "The AR-VR age has begun in health care." November 2020. [Online]. Available: https://www.cta.tech/Resources/i3-Magazine/i3-Issues/2020/November-December/The-AR-VR-Age-has-Begun-inHealth-Care.

[20] "AR and VR in education market." June 2022. [Online]. Available: https://www.market-researchfuture.com/reports/ar-vr-in-educationmarket-10834.

[21] J. Thomason, "MetaHealth-how will the metaverse change health care?" *Journal of Metaverse*, vol. 1, no. 1, pp. 13–16, 2021.

[22] B. Kye, N. Han, E. Kim, Y. Park, and S. Jo, "Educational applications of metaverse: Possibilities and limitations." *Journal of Educational Evaluation for Health Professions*, vol. 18, p. 32, 2021.

[23] T. Ruohomäki, E. Airaksinen, P. Huuska, O. Kesäniemi, M. Martikka, and J. Suomisto, "Smart city platform enabling digital twin." In: Proceedings of the International Conference on Intelligent Systems (IS), Funchal, Portugal, 2018, pp. 155–161.

[24] J. Y. Lee, "A study on metaverse hype for sustainable growth." *International Journal of Advanced Smart Convergence*, vol. 10, no. 3, pp. 72–80, 2021.

[25] M.S. Anwar, J. Wang, W. Khan, A. Ullah, S. Ahmad, and Z. Fei, "Subjective QoE of 360-degree virtual reality videos and machine learning predictions." *IEEE Access*, vol. 8, pp. 148084–148099, 2020.

[26] M.S. Anwar, J. Wang, A. Ullah, W. Khan, S. Ahmad, and Z. Fei, "Measuring quality of experience for 360-degree videos in virtual reality." *Science China Information Sciences*, vol. 63, pp. 1–15, 2020.

[27] M. Shahid Anwar, J. Wang, S. Ahmad, A. Ullah, W. Khan, and Z. Fei, "Evaluating the factors affecting QoE of 360-degree videos and cybersickness levels predictions in virtual reality." *Electronics*, vol. 9, no. 9, p. 1530, 2020.

[28] M.S. Anwar, J. Wang, A. Ullah, W. Khan, Z. Li, and S. Ahmad, "User profile analysis for enhancing QoE of 360 panoramic videos in virtual reality environment." In: *2018 International Conference on Virtual Reality and Visualization (ICVRV)*, Qingdao, China, 2018, pp. 106–111. IEEE.

[29] M.S. Anwar, J. Wang, A. Ullah, W. Khan, S. Ahmad, and Z. Li, "Impact of stalling on QoE for 360-degree virtual reality videos." In: *2019 IEEE International Conference on Signal, Information and Data Processing (ICSIDP)*, Chongqing, China, 2019, pp. 1–6. IEEE.

[30] M.S. Anwar, J. Wang, S. Ahmad, W. Khan, A. Ullah, M. Shah, and Z. Fei, "Impact of the impairment in 360-degree videos on users' VR involvement and machine learning-based QoE predictions." *IEEE Access*, vol. 8, pp. 204585–204596, 2020.

[31] K. G. Nalbant and S. Uyanik, "Computer vision in the metaverse." *Journal of Metaverse*, vol. 1, no. 1, pp. 9–12, 2021.

[32] H. Ning et al., "A survey on metaverse: The state-of-the-art, technologies, applications, and challenges." arXiv:2111.09673, 2021.

[33] H. B. Demeke et al., "Trends in use of telehealth among health centers during the COVID-19 pandemic-United States, June 26-November 6, 2020." *Morbidity and Mortality Weekly Report*, vol. 70, no. 7, p. 240, 2021.

[34] A. Musamih et al., "Metaverse in healthcare: Applications, challenges, and future directions." *IEEE Consumer Electronics Magazine*, vol. 12, no. 4, pp. 33–46, 2022. https://doi.org/10.1109/MCE.2022.3223522.

[35] K. Li et al., "An experimental study for tracking crowd in smart cities." *IEEE Systems Journal*, vol. 13, no. 3, pp. 2966–2977, 2019.

[36] B. Wei, K. Li, C. Luo, W. Xu, J. Zhang, and K. Zhang, "No need of data pre-processing: A general framework for radio-based device free context awareness." *ACM Transactions on Internet of Things*, vol. 2, no. 4, pp. 1–26, 2021.

[37] B. P. L. Lau, T. Chaturvedi, B. K. K. Ng, K. Li, M. S. Hasala, and C. Yuen, "Spatial and temporal analysis of urban space utilization with renewable wireless sensor network." In: *Proceedings of the 3rd IEEE/ACM International Conference on Big Data Computing, Applications and Technologies*, New York, USA, 2016, pp. 133–142.

[38] K. Li, C. Yuen, and S. Kanhere, "Senseflow: An experimental study of people tracking." In: *Proceedings of the 6th ACM Workshop on Real World Wireless Sensor Networks*, New York, USA, 2015, pp. 31–34.

[39] M. G. Hanna, I. Ahmed, J. Nine, S. Prajapati, and L. Pantanowitz, "Augmented reality technology using microsoft HoloLens in anatomic pathology." *Archives of Pathology & Laboratory Medicine*, vol. 142, no. 5, pp. 638–644, 2018.

[40] K. Yue, "Breaking down the barrier between teachers and students by using metaverse technology in education: Based on a survey and analysis of Shenzhen City, China." In: *Proceedings of the 13th International Conference on E-Education, E-Business, E-Management, and E-Learning (IC4E 2022)*, Tokyo, Japan, 14–17 January 2022, pp. 40–44.

[41] Z. Li, J. Wang, Z. Yan, X. Wang, and M. S. Anwar, "An interactive virtual training system for assembly and disassembly based on precedence constraints." In: *Advances in Computer Graphics: 36th Computer Graphics International Conference, CGI 2019*, Calgary, AB, Canada, 17–20 June 2019, Proceedings 36, 2019, pp. 81–93. Springer International Publishing.

[42] Z. Li, J. Wang, M. S. Anwar, and Z. Zheng, "An efficient method for generating assembly precedence constraints on 3D models based on a block sequence structure." *Computer-Aided Design*, vol. 118, p. 102773, 2020.

[43] Z. Li, S. Zhang, M. S. Anwar, and J. Wang, "Applicability analysis on three interaction paradigms in immersive vr environment." *In: 2018 International Conference on Virtual Reality and Visualization (ICVRV)*, Qingdao, China, 2018, pp. 82–85. IEEE.

[44] N. Rajagopal, J. Miller, K. K. R. Kumar, A. Luong, and A. Rowe, "Improving augmented reality relocalization using beacons and magnetic field maps." In: *Proceedings of the International Conference on Indoor Positioning and Indoor Navigation (IPIN)*, Pisa, Italy, 2019, pp. 1–8.

[45] S. Han and L. Ci, "Research trends on augmented reality education in Korea from 2008 to 2019." *Journal of Educational Technology*, vol. 36, pp. 505–528, 2020.

[46] A. Tlili, R. Huang, B. Shehata, D. Liu, A. Hosny, S. Metwally, H. Wang, M. Denden, A. Bozkurt, L.-H. Lee et al., "Is metaverse in education a blessing or a curse: A combined content and bibliometric analysis." *Smart Learning Environments*, vol. 9, p. 24, 2022.

[47] C. Meier, J. Saorín, A. B. de León, and A. G. Cobos, "Using the roblox video game engine for creating virtual tours and learning about the sculptural heritage." *International Journal of Emerging Technologies in Learning*, vol. 15, no. 20, pp. 268–280, 2020.

[48] B. Ryskeldiev, Y. Ochiai, M. Cohen, and J. Herder, "Distributed metaverse: Creating decentralized blockchain-based model for peer-to-peer sharing of virtual spaces for mixed reality applications." In: *Proceedings of the 9th Augmented Human International Conference*, New York, USA, 2018, p. 39. [Online]. Available: https://doi.org/10.1145/3174910.3174952.

[49] H. C. Hou and H. Wu, "Technology for real estate education and practice: A VR technology perspective." *Property Management*, vol. 38, no. 2, pp. 311–324, 2020.

[50] M. Izani, S. Aalkhalidi, A. Razak, and S. Ibrahim, "Economical VR/AR method for interior design programme." In: *Proceedings of the Advances in Science and Engineering Technology International Conference (ASET)*, Dubai, United Arab Emirates, February 2022, pp. 1–5.

[51] C. Aiello, J. Bai, J. Schmidt, and Y. Vilchynskyi, "Probing reality and myth in the metaverse." McKinsey & Company, 2022. Retrieved from: https://www.mckinsey.com/industries/retail/our-insights/probing-reality-and-myth-in-the-metaverse.

[52] WGSN Insight Team, Luxury brand strategies 2022. WGSN, 2022. Retrieved from: https://www-wgsn-com.ezproxy.lb.polyu.edu.hk/insight/article/93198#page2.

[53] M. Blázquez, "Fashion shopping in multichannel retail: The role of technology in enhancing the customer experience." *International Journal of Electronic Commerce*, vol. 18, no. 4, pp. 97–116, 2014.

[54] J.-G. Kim, "A study on metaverse culture contents matching platform." *The International Journal of Advanced Culture Technology*, vol. 9, no. 3, pp. 232–237, 2021.

[55] T. Lyttelton, E. Zang, and K. Musick, "Telecommuting and gender inequalities in parents' paid and unpaid work before and during the COVID-19 pandemic." *Journal of Marriage and the Family*, vol. 84, no. 1, pp. 230–249, 2022.

[56] K. Li, W. Ni, B. Wei, and M. Guizani, "An experimental study of twoway ranging optimization in UWB-based simultaneous localization and wall-mapping systems." In: *Proceedings of the International Wireless Communications and Mobile Computing (IWCMC)*, Dubrovnik, Croatia, 2022, pp. 871–876.

[57] K. Li, W. Ni, and P. Zhang, "Poster: An experimental localizationtestbed based on UWB channel impulse response measurements." In: *Proceedings of the* 21st ACM/IEEE International Conference on Information Processing in Sensor Networks (IPSN), Milano, Italy, 2022, pp. 515–516.

[58] F. Tao, H. Zhang, A. Liu, and A. Y. C. Nee, "Digital twin in industry: State-of-the-art." *IEEE Transactions on Industrial Informatics*, vol. 15, no. 4, pp. 2405–2415, 2019.

[59] D. Chen, D. Wang, Y. Zhu, and Z. Han, "Digital twin for federated analytics using a Bayesian approach." *IEEE Internet of Things Journal*, vol. 8, no. 22, pp. 16301–16312, 2021.

[60] M. M. Rathore, S. A. Shah, D. Shukla, E. Bentafat, and S. Bakiras, "The role of AI, machine learning, and big data in digital twinning: A systematic literature review, challenges, and opportunities." *IEEE Access*, vol. 9, pp. 32030–32052, 2021.

[61] D. Gelernter, *Mirror Worlds: Or the Day Software Puts the Universe in a Shoebox...How It Will Happen and What It Will Mean*. Oxford, UK: Oxford University Press, 1993.

[62] Y. K. Dwivedi et al., "Metaverse beyond the hype: Multidisciplinary perspectives on emerging challenges, opportunities, and agenda for research, practice and policy." *International Journal of Information Management*, vol. 66, 2022, Art. no. 102542.

[63] A. Kuppa and N.-A. Le-Khac, "Black box attacks on explainable artificial intelligence (XAI) methods in cyber security." In: *Proceedings of the International Joint Conference on Neural Networks (IJCNN)*, Glasgow, UK, 2020, pp. 1–8.

[64] "Metaverse security threats in the future of internet." [Online]. Available: https://www.juegostudio.com/blog/Metaverse-security-threats-in-the-future-of-internet [Accessed on: February 22, 2023].

[65] S. Qamar, Z. Anwar, and M. Afzal, "A systematic threat analysis and defense strategies for the metaverse and extended reality systems." *Computers Security*, vol. 128, p. 103127, 2023.

[66] T. Oleksy, A. Wnuk, and M. Piskorska, "Migration to the metaverse and its predictors: Attachment to virtual places and metaverse-related threat." *Computers in Human Behavior*, vol. 141, p. 107642, 2023.

[67] S.-Y. Kuo, F.-H. Tseng, and Y.-H. Chou, "Metaverse intrusion detection of wormhole attacks based on a novel statistical mechanism." *Future Generation Computer Systems*, vol. 143, pp. 179–190, 2023.

[68] A. K. Ghosh, A. S. Ullah, R. Teti, and A. Kubo, "Developing sensor signal-based digital twins for intelligent machine tools." *Journal of Industrial Information Integration*, vol. 24, 2021, Art. no. 100242.

[[69] K. Li, W. Ni, E. Tovar, and M. Guizani, "Optimal rate-adaptive data dissemination in vehicular platoons." *IEEE Transactions on Intelligent Transportation Systems*, vol. 21, no. 10, pp. 4241–4251, 2020.

[70] R. S. Kenett and J. Bortman, "The digital twin in industry 4.0: A wideangle perspective." *Quality and Reliability Engineering International*, vol. 38, no. 3, pp. 1357–1366, 2022.

[71] K. Li, L. Lu, W. Ni, E. Tovar, and M. Guizani, "Secret key agreement for data dissemination in vehicular platoons." *IEEE Transactions on Vehicular Technology*, vol. 68, no. 9, pp. 9060–9073, 2019.

[72] S. Boschert and R. Rosen, "Digital twin-The simulation aspect." In: *Mechatronic Futures: Challenges and Solutions for Mechatronic Systems and Their Designers*, P. Hehenberger and D. Bradley, Eds. Cham, Switzerland: Springer Int., 2016, pp. 59–74.

[73] K. Li, R. C. Voicu, S. S. Kanhere, W. Ni, and E. Tovar, "Energy efficient legitimate wireless surveillance of UAV communications." *IEEE Transactions on Vehicular Technology*, vol. 68, no. 3, pp. 2283–2293, 2019.

[74] Y. Jiang, S. Yin, K. Li, H. Luo, and O. Kaynak, "Industrial applications of digital twins." *Philosophical Transactions of the Royal Society A: Mathematical, Physical and Engineering Sciences*, vol. 379, no. 2207, 2021, Art. no. 20200360.

[75] K. Li, W. Ni, M. Abolhasan, and E. Tovar, "Reinforcement learning for scheduling wireless powered sensor communications." *IEEE Transactions on Green Communications and Networking*, vol. 3, no. 2, pp. 264–274, 2019.

[76] E. U. Haque, W. Abbasi, S. Murugesan, M. S. Anwar, F. Khan, and Y. Lee, "Cyber forensic investigation infrastructure of Pakistan: An analysis of cyber threat landscape and readiness." *IEEE Access*, vol. 11, pp. 40049–40063, 2023.

[77] K. Li et al., "Fair scheduling for data collection in mobile sensor networks with energy harvesting." *IEEE Transactions on Mobile Computing*, vol. 18, no. 6, pp. 1274–1287, 2019.

[78] Z. Liu, N. Meyendorf, and N. Mrad, "The role of data fusion in predictive maintenance using digital twin." In: *Proceedings of the AIP Conference*, Provo, Utah, USA, 2018, Art. no. 20023. [Online]. Available: https://aip.scitation.org/doi/abs/10.1063/1.5031520.

[79] K. Li, N. Lu, J. Zheng, P. Zhang, W. Ni, and E. Tovar, "Bloothair: A secure aerial relay system using Bluetooth connected autonomous drones." *ACM Transactions on Cyber-Physical Systems*, vol. 5, no. 3, pp. 1–22, 2021.

[80] K. Li et al., "Design and implementation of secret key agreement for platoon-based vehicular cyber-physical systems." *ACM Transactions on Cyber-Physical Systems*, vol. 4, no. 2, pp. 1–20, 2019.

[81] D.-I. D. Han, Y. Bergs, and N. Moorhouse, "Virtual reality consumer experience escapes: Preparing for the metaverse." *Virtual Reality*, vol. 26, pp. 1443–1458, 2022.

[82] S. Mystakidis, "Metaverse." *Encyclopedia*, vol. 2, no. 1, pp. 486–497, 2022.

[83] M. A. I. Mozumder, M. M. Sheeraz, A. Athar, S. Aich, and H.-C. Kim, "Overview: Technology roadmap of the future trend of Metaverse based on IoT, blockchain, AI technique, and medical domain metaverse activity." In: *Proceedings of the International Conference on Advanced Communication Technology (ICACT)*, PyeongChang Kwangwoon Do, Korea, 2022, pp. 256–261.

[84] T. R. Gadekallu et al., "Blockchain for the metaverse: A review." *arXiv:2203.09738*, 2022.

Social Internet of Things (SIoT)

Recent Trends and Its Applications

Irshad Khalil
Gachon University

Adnan Khalil
University of Malakand

Inam Ullah
Gachon University

Yuning Tao
South China University of Technology

Ijaz Khan
Harbin Institute of Technology

Shahzad Ashraf
NFC Institute of Engineering and Technology

Waleed M. Ismael
Azal University for Human Development

DOI: 10.1201/9781032648309-12

9.1 INTRODUCTION

Kevin Ashton was the first person who introduced the idea of Internet of Things (IoT) and proposed the idea of linking Radio Frequency Identification (RFID) to the internet [1]. By using this approach, each physical object becomes a smart object when it is connected to the internet, i.e., wall, door, and other appliances. These technologies may include different sensors, wireless, and web technologies. These devices have different features such as architecture, operating system, platform, and communication protocols, and they are connected to each other to exchange data and fulfill the need of the end users [2]. According to the forecast of Cisco [3] and Garter [4], the number of these smart devices may increase from 20 to 50 billion in 2022, which greatly increases the IoT network complexities [5]. Managing such a huge network of billions of devices is very difficult because of the lack of universal standards, and the current IoT standards and protocols are not cost-effective and avoid the realization of IoT's true potential [6].

In recent years, research has been carried out, which shows that the concept behind Social Networks (SN) and IoT show underlying similarities [7], and different research efforts are made on the integration of SN in the IoT world to solve different implementation problems [8–11]. The Social Internet of Things (SIoT) is a newly described term in computer network, which relates the concept of SN to the IoT. Reference [12] proposed a new approach based on the concept of relationships in social objects in IoT networks without human intervention. Using this approach, smart objects become social objects when social value or properties are added to the smart objects. These social properties enable the social objects to build/establish social relationship autonomously with other social objects to carry out different tasks. These tasks may form simple (using smart mobile) to very complex tasks sharing the very complicated infrastructure of the city. SIoT uses all the devices connected from anywhere to create relationships based on some common interest and provide better services to the end users [13]. Different types of relationships are established among different social objects, which include parental object, colocation, co-work, and social object relation to fulfill different types of needs according to user needs and requirements. In the SIoT network, social objects communicate like social agents to exchange information against generated queries [14]. Social interaction in the SIoT was inspired from the well-known statistical theory called Fiske theory [15], which shows the real communication pattern among human beings and the types of relationships in communities.

The main objective of enabling SN in IoT is to allow users to set the rules for social objects to protect the privacy of the smart/social device and only access the throughput/results of inter-object interaction [16]. Enabling the SN for smart objects can improve different network parameters such as service discovery, trustworthiness, friendship selection, and network navigability [12] and suggested that by having an SN capability for social objects, one can efficiently achieve network navigability, friendship management, interoperability in a trustworthy environment, and service discovery efficiently.

Using the concept of SN and IoT greatly improves and explains well the complexity of interaction among users and social things in the network [17]. The concept of SIoT started in 2011, and still now, many efforts are being made to improve the different concepts of SIoT [18]. The concept of SIoT is at its earlier stage, and effort has been made to propose the idea of integrating SIoT into actual application [19,20]. Despite the increasing popularity of this active research area, there are very few related research studies that have reviewed different notions of SN in SIoT (e.g., [12,18,21,22]).

9.2 RELATED LITERATURE REVIEWS

Related internet-enabling technologies including ontologies [23], machine learning (ML) [24], deep learning (DL) [25], human-computer interfaces (HCI) [26], and other relevant technologies are greatly contributing to the adaptation of SIoT. In the recent past, many research efforts have been made to apply these approaches in different aspects of SIoT. Several survey papers have been recently published focusing on different SIoT aspects, i.e., network navigation techniques, trustworthiness, friendship selection, and relationship management techniques. These papers introduced a comprehensive overview of these aspects. To the best of our knowledge, until now, there is no review paper that focuses on the application of SIoT in the application domain (applications where the concept of SIoT was fully implemented). In this chapter, we provide overviews of concepts, relationship, platform and implementation, architecture, and application area where the concept of social IoT is used.

The main objective and contributions of this chapter are to:

- Offer an in-depth understanding of the fundamentals of the SIoT domain, including different aspects and types of relationships within SIoT.

- Review various platforms that support the implementation and realization of the SIoT concept.

- Examine various architectures and datasets that enable the development of SIoT applications.

- Provide a comprehensive examination of different cloud-based SIoT applications.

- Summarize the various application domains where the SIoT concept has been implemented and applied.

9.3 METHOD FOR PAPERS COLLECTION

In this review chapter, we have used the PRISMA approach for the collection and analysis of recent research in SIoT-related areas. The review process included three steps: paper identification, selection, and analysis. To identify relevant papers, we searched major online research repositories such as IEEE Xplore, SpringerLink, ScienceDirect, MDPI, Hindawi, the ACM Digital Library, and Google Scholar as these repositories are much related to the computer science domain. Additionally, we also included publications from Web of Science, Scopus, and Science Direct to capture the multidisciplinary nature of the topic. We collected all related papers that matched the search criteria and further screened them for inclusion in the review.

In the second step, we scanned the collected research paper records and removed the nonconforming and non-relevant papers. Finally, after this stage, only papers relevant to the concept of SIoT and its application are included in our critical review. Figure 9.1 shows the proposed method for this review.

After a thorough review and selection process, we only included papers that were relevant to the concept of either SIoT or Social Web of Things. All other papers were excluded from this review. We considered articles that proposed theoretical solutions, conceptual approaches, architectures, frameworks for integrating SN and IoT in SIoT/SWoT, and real-world applications of the SIoT concept, as well as partially or fully completed prototypes or applications using the SIoT concept.

The total number of papers that matched the exact term "Social Internet of Things" in the different scientific libraries mentioned previously were considered for further evaluation

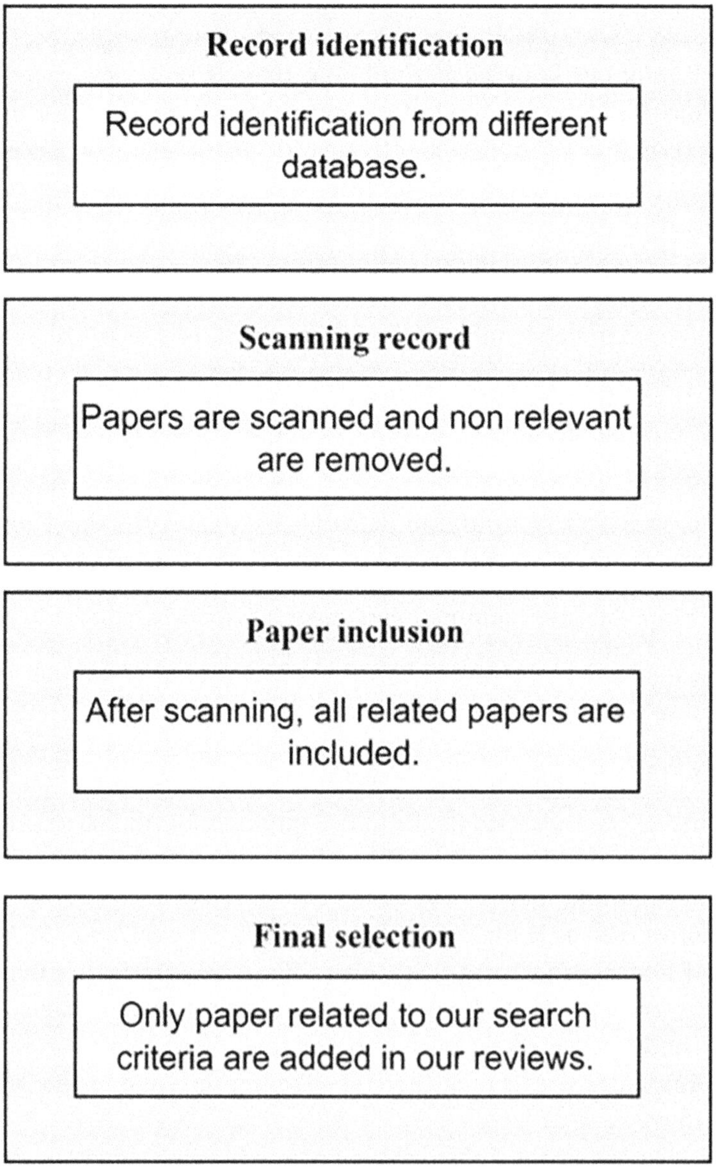

Record identification

Record identification from different database.

Scanning record

Papers are scanned and non relevant are removed.

Paper inclusion

After scanning, all related papers are included.

Final selection

Only paper related to our search criteria are added in our reviews.

FIGURE 9.1 PRISMA study selection diagram.

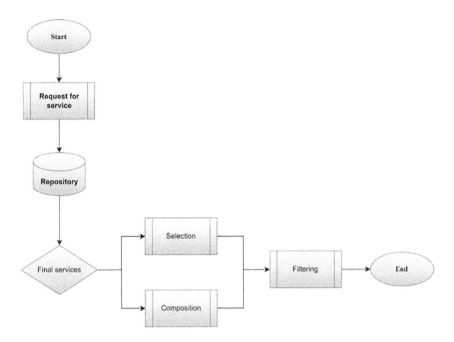

FIGURE 9.2 Number of papers matching the exact term "Social Internet of Things."

and analysis. Figure 9.2 illustrates the number of papers in each repository that exactly matched the term "Social Internet of Things."

These papers are analyzed for their relationship, platform, and different implementations, as well as cloud-based applications and real-world scenarios where the concept of SIoT is fully implemented.

9.4 BACKGROUND

In recent years, the idea of converging the concept of SN with IoT has gained more importance. Kleinberg [27] first introduced the concept of converging the idea of SN with IoT applications through which smart objects are capable of creating a social relationship with each other to fulfill a common goal. The most important step in the direction of SIoT has been considered in Ref. [26]. In this paper, the integration between SN and IoT and how SN is used to bring social aspects in smart objects are discussed. The author does not propose the required architecture for SIoT and does not describe how social relationships are established among social objects. These social attributes are explained in Ref. [28], which enable social interaction among social objects. The relationship model for SIoT was inspired from the well-known Fiske theory [15], which presents a complete relationship model for human society. These objects can communicate with each other and behave like social agents in the social network. These social agents can request and respond from/to other social agents to provide social services [14]. Each social object can directly communicate with each other to improve and with humans based on a set of rules set by humans such as how to select friends and make friendship or how to provide services [12]. The most recent advancement in SIoT Social Collaborative smart thing [29] is by which social objects collaborate and create social relationship with others. For the readers, Refs. [30] and [16] present the complete evaluation history of how a smart object becomes a social object. In SIoT, different relations are created, which provide service in a distributed manner rather than relying on simple server-client architectures [29]. Unique benefits of using the concept of SIoT are as follows:

1. Whole SIoT network is navigable.

2. Improved scalability.

3. Objects can establish relationships with each other without human involvement [12].

4. SIoT increases security because services are available from friend object.

9.5 ARCHITECTURE OF SIoT

There is no standard architecture for different architectures. Most of the research articles rely on four-layer architectures. These layers include the object layer (consists of a collection of objects and sensors), global connection layer, platform layer, and application layer (presenting information to the end user) [31]. This is a perfect architecture for an IoT network and cannot be used for the development of SIoT. In SIoT, these four layers are similar to the standard IoT architecture, and there is a need for the fifth layer, which brings social interaction among social objects. Almost there is a standard SIoT architecture having same layers, but these architectures are modified or new components are added for the development of different applications.

Recent contributions regarding the development of SIoT architectures are presented in Table 9.1 along with layer, structure details, domains, and detailed descriptions.

Atzori et al. [12] proposed a three-layered architecture for SIoT. The proposed architecture consists of the following layers:

- **Object Layer:** This layer comprises of physical devices and their communication interfaces. It represents the physical entities that are connected to the network and can be represented by an object identifier. The object layer is responsible for the management of the physical devices, their location, and their current state.

- **Component Layer:** The component layer is responsible for managing different components of the SIoT system, including Object Profiling, Identification Management, Owner Control, Relationship Management, Service Discovery, and Composition components. The Object Profiling component is responsible for managing the characteristics of the objects and their description. The Identification Management component is responsible for managing the unique identification of the objects. The Owner Control (OC) component is responsible for managing the ownership of the objects. The Relationship Management Component (MC) is responsible for managing the relationships between the objects. The Service Discovery component is responsible for discovering the services offered by the objects. The Composition component is responsible for managing the interactions between the objects. Additionally, there

TABLE 9.1   Architecture Details of SIoT

Article	Domain	Architecture	Description
Atzori et al. [12]	SIoT	Three layers	Architecture based on the socialized object concepts
Atzori et al. [32]	SIoT	Three layers	This is a slight variation in which three layers are combined in a single layer
Zhang et al. [33]	SIoT	Three-layer	Language processing and machine learning were used for processing
Byun et al. [34]	Ontology-based SIoT model	Three-layer	Integration of SN with IoT with a purpose to create a relationship among another social object
Kim et al. [35]	SIoT	Rational structure	Proposed a socialite architecture that combines social objects
Alam et al. [36]	SIOT (SIoVT)	Six layers	Application of social graph for friendship representation
Talal et al. [37]	SIoT (SIoVT)	Three-layer	Scalable architecture based on restful web technologies
Farhan Amin et al. [38]	SIoT	Four layers	Capable of selecting friend and searching for required services
Gulati et al. [19]	SIoT	Four layers	Semantic-oriented architecture for SIoT
Dutta et al. [30]	SIoT	Layered	Client-server architecture for SIoT
Voutyras et al. [39]	SIoT	Rational structure	Architecture based on four different group
Kosmatos et al. [40]	SIoT	Four layer	Semantic and server-oriented architecture
Voutyras et al. [41]	SIoT	Layer architecture	Four different groups were used for all SIoT activities

is a component for trustworthiness management, which is responsible for managing the trustworthiness of the objects.

- **Application Layer:** The application layer consists of an interface for users, objects, and different services API for communication. It enables the users to interact with the objects and services and provides a way for the objects and services to interact with each other. The application layer provides an API for the communication between the objects and services and provides an interface for the users to interact with other social objects and services. Overall, the three-layered architecture proposed by Atzori et al. [12] provides a clear and structured way to manage the different components of the SIoT system and enables the efficient communication and interaction between the objects, services, and users. In Ref. [32], a slight variation of the architecture proposed by Atzori et al. [12] is the combination of the three layers into a single component called the SIoT Server layer, which includes the Application and Network layers. This layer is composed of three sub-layers:

- **SIoT Server Layer:** This layer is responsible for managing the different components of the SIoT system, including Object Profiling, Identification Management, Owner Control, Relationship Management, Service Discovery, and Composition components. It also includes a component for trustworthiness management.

- **Gateway Layer:** This layer is responsible for providing a bridge between the SIoT server and the objects, enabling communication and interaction between them. This layer is responsible for managing the communication protocols between the objects and the SIoT server and for managing the data flow between them.

- **Object Layer:** This layer comprises of physical devices and their communication interfaces. It represents the physical entities that are connected to the network and can be represented by an object identifier. The object layer is responsible for the management of physical devices, their location, and their current state. This variation of the architecture combines the three layers into a single component, the SIoT Server layer, which simplifies the management of the different components of the SIoT system. The SIoT Server layer provides an integrated solution for managing the different components of the SIoT system and enables efficient communication and interaction between the objects, services, and users. The gateway layer is responsible for managing the communication protocols between the objects and the SIoT server, and for managing the data flow between them. Moreover, the object layer represents the physical entities that are connected to the network and can be represented by an object identifier. In Ref. [33], another architecture for SIoT is presented based on the Web of Things (WoT) and SN. This architecture utilizes machine learning for processing and interpreting natural languages into machine language, making the devices understandable by human beings. The status of the devices is then shared with other devices through social networks.

Communication between devices is enabled using Restful APIs. This proposed Social Web of Things architecture is composed of three layers:

- **External Resource Layer:** This layer consists of any smart devices that can connect to the internet through Restful APIs. These devices include sensors, actuators, and other third-party vendor sensors.

- **Platform Layer:** This layer is the core of the proposed architecture and includes all functionalities such as the processing of natural languages to machine languages, interpretation of natural languages, identifying resources, and Business Process Management tasks.

- **Application Layer:** This layer is responsible for presenting the information to the user and other third-party applications. It provides information to the end users. The Lilliput architecture, as presented by Byun et al. [34], is a system designed to manage and interact with IoT devices. It comprises several functional components, including the following: The Reflection Manager, which is responsible for receiving and storing online information about entities in a social graph; the IoT Social Graph Manager, which builds and maintains a social graph of IoT components and directly interacts with four different sub-components; the Entity Manager, which manages the IDs of IoT devices and verifies their class or object type; the Relationship Manager, which manages all the relationships that occur within the system; the Synchronization Manager, which maintains a list of callback receivers; the Change Notification Manager, which notifies users when changes occur in the social graph; the Graph Utilization Manager, which receives queries from users and returns results; the Modification Manager, which modifies the IoT social graph using knowledge base APIs; the Reflection Manager, which performs bi-directional reflection between cloud space and IoT social networks; the Invoker component, which invokes different smart services on social things and displays results for the application; and the Security Manager, which determines whether requests from applications are suitable or not. The Lilliput architecture also provides RESTful APIs for interaction with the system.

Kim et al. [35] proposed a three-layer architecture for the SIoT called Socialite. This architecture is designed to provide a flexible and scalable solution for managing and interacting with IoT devices in a social context. The first layer is the Socialite client application layer, which is reliable for providing access to the different devices and services in the system. This layer is also responsible for programming the rules and communication mechanisms that govern the interactions between devices and users. The client application layer can be programmed using a variety of programming languages, providing flexibility and ease of use for developers. The second layer is the Socialite Server layer, which acts as a gateway between the client application layer and the different types of devices and services in the system. This layer is responsible for providing access to devices of different architectures and protocols and for managing the interactions between devices and the client application layer. Finally, at the third layer, the database layer, all the information about the devices and their relationships is stored. This layer allows for efficient storage and retrieval of data and is designed to handle large amounts of data and the scalability of the system. The database layer also helps in maintaining the state of the devices and the relationships that have been established between them. Overall, the Socialite architecture is designed to provide a flexible and scalable solution for managing and interacting with IoT devices in a social context by providing a three-layer architecture that separates the different functionalities of the system.

Alam et al. [36] proposed a six-layered architecture for the SIoVT. The architecture is designed to provide a comprehensive solution for managing and interacting with connected vehicles in a social context. The proposed architecture consists of six layers, each with specific functions and responsibilities. The first layer is the Home Base Unit (HBU) layer, which is responsible for managing data, handling the dispatching of messages, and setting privacy settings. This layer is the primary point of interaction between the vehicle and the user. It includes a Data Manager, Dispatcher, and Privacy settings. The second layer is the On-Board Unit (OBU) layer, which is responsible for managing the identity of the vehicle, building messages, managing data, and handling dispatching. This layer includes an Identity Manager, Message Builder, Data Manager, and Dispatcher. The third layer is the tNote Message layer, which is responsible for handling dedicated Short Range Communications and Advanced Traveler Information System (ATIS) for the vehicle. This layer is responsible for sending and receiving messages between vehicles and other road users. The fourth layer is the Road Side Unit (RSU) layer, which is responsible for managing the identity of the vehicle, handling data management, dispatching messages, and managing social tags. This layer includes an Identity Manager, Data Manager, Dispatcher, and Social Tag Manager. The fifth layer is the tNote Cloud layer, which is responsible for handling the topology optimization, query processing, data management, and providing a user interface for managing routes, friends, and groups and generating social graphs. This layer includes a Topology Optimizer, Query Processor, Data Manager, and User Interface. Lastly, the sixth layer is the User Interface layer, which is responsible for providing a user-friendly interface for managing user profiles, routes, friends, and groups and generating social graphs. This layer includes features for managing routes, friends, and groups and generating social graphs. Overall, the proposed six-layered architecture for SIoVT provides a comprehensive solution for managing and interacting with connected vehicles in a social context, by providing a structured and organized way to handle the different functionalities of the system. Reference [37] proposed a four-layered architecture for implementing the concept of SIoT in the IoV domain. The proposed architecture also consists of six layers that include the following: (1) Physical world layer, which consists of real-world objects, i.e., cars and other transportation vehicles; (2) Gateway layer, which consists of smart vehicles and other roadside units; (3) Fog layer, which is used for the management of Fog nodes components in the SIoV networks; (4) Cloud layer, which is responsible for the big data, resource, and analytic; (5) Application layer, which is used for application management and services; and (6) User layer users, pedestrians, and Intellection traffic system units. Reference [38] proposed an architecture that implements both the functionality of IoT and SIoT. There are four components in the Social Pal platform, which are as follows: (1) Actor is any device or person who interacts with the system; (2) Social Pal is responsible for the discovery of the service for the social objects; (3) Interface is responsible for making new connection for social objects; and (4) Internet provides a way of access to each component in the platform. This SIoT platform has inherent important features of friendship management, services decomposition, and others from social networks.

Gulati et al. [19] proposed a semantic-oriented architecture model as reference for SIoT. The proposed architecture consists of four layers. Object layer is referred to the collection of all social objects that may be included in the network. Communication layer is responsible for communication among all social objects. Social management layer is responsible for managing the relationship and assigning ID to each object. Presentation layer aims to provide information to the end users and consists of mobile and web-based applications. Voutyras et al. [39] proposed an SIoT architectural model, which is quiet similar to Ref. [12] except two new modules, namely, Mobile object relationship and Explorer object relationship, which are added to the main component. The functionality of the proposed architecture is the same as in Ref. [12]. The EoR module is responsible for creating connections with movable and static objects, while the MoR module is used to establish the relationship with smart objects that travel among them. Reference [40] proposed a unified architecture for the IoT that integrates smart objects and RFID devices to create a social network and explore a social feature of the smart object. In this architecture, smart objects create a connection with other smart objects and create a social relationship to deliver services to other objects. However, the proposed unified architecture in not implemented. Reference [41] proposed an SIoT architecture based on a relational model including four different groups as a basic design element. These smart objects are integrated with SN and other social objects to form relationships using the module of COSMOS management frameworks.

9.6 RELATIONSHIP IN SIoT

In SIoT, smart objects can create relationships with other objects. Different research articles focus on the types of relationships among other objects. Atzori et al. [12] proposed five different types of relationships among social objects. Roopa et al. [42] classified these types into further two types of relationships, i.e., object-to-object and human-to-object relationships. To develop an SIoT application, selecting appropriate types of relationships plays an important role. Reference [12] presented different applications of the SIoT using different types of social relationships based on the social objects. Table 9.1 describes different application domains based on these relationships. In the following sub-section, we briefly discussed different types of relationships that are established among different social objects in an SIoT environment. These relationships are created among either user to object or among object to object in an SIoT domain.

1. **UO Relationship:**

   - **Ownership Object Relationship (OOR):** OOR is created among objects of the same user, i.e., personal laptop, personal mobiles, smart car, and all other smart objects belonging to the same user.

   - **Social Object Relationship (SOR):** SOR belongs to friends, i.e., interchange of phone numbers when friends are in contact with others [43].

   - **Sibling Object Relationship (SiOR):** This relationship is created among different smart objects that belong to a family member [33].

   - **Guest Object Relationship (GOR):** This relationship is created by smart objects owned by the users as a guest in a specific role.

2. **OO Relationship:**

   - **Parental Object Relationship (POR):** This relationship is created among similar objects of the same company or distributor.

   - **Co-location Object Relationship:** This relationship is established among objects in the same location.

   - **Co-work Object Relationship:** This relationship is established by smart objects that work together to provide service for a common IoT application.

   - **Guardian Object Relationship (GoR):** This relationship is established when social vehicles turn into child objects in association with the super objects of Road Side Units [34].

   - **Stranger Object Relationship:** This relationship is established among objects in a public gathering.

   - **Service Object Relationship:** This type of relationship is established among smart objects that fulfill the required services that are requested by coordinating the same service composition.

TABLE 9.2   Types of Relationship and Possible Application

	Object to Object Relationship	
	POR	1. Smart printer 2. Smart card reading 3. Personal laptop
	CLOR	1. Smart hospital 2. Smart parking 3. Smart office 4. Smart office
Object-to-object relationship	CWOR	1. Telemedicine 2. Remote patient monitoring 3. Emergency response system 4. On demand doctors
	GOR	1. Early warning system 2. Tour application (road blocking)
	STGOR	1. Smart marketing 2. Tailored customization 3. Campaign management
	SVOR	1. Smart museum application 2. Location-based services 3. Recommendation services 4. Detect unusual situation
	Human-Object Relationship	
	SOR	1. E toll collection 2. Lane change assistance
	OOR	1. Smart energy management 2. Smart transportation 3. Smart logistics
Human-object relationship	SIBOR	1. Smart stadium 2. Game statistic 3. Crowd management
	GSTOR	1. Smart restaurant 2. Smart shopping 3. Smart bill payment

Table 9.2 displayed the list of different types of possible relationships.

9.7 PLATFORM, IMPLEMENTATION, AND DATASET

There are different platforms that enable easier and reliable interactions among social objects to achieve a common goal. In this section, we studied the recent research contribution and efforts to design and develop such platforms. Several projects have been created with the aim of integrating IoT and social networks.

Toyota friend platform is one of the earliest platforms, which made the data generated by smart objects available on social networks. It is a private network that aims to collect data among automobiles and create a community to increase customer satisfaction. *Nike +* is another platform that collects data from the sensors for Nike shoes and then puts these data on social networks. This is solely a private project for customer satisfaction, and no APIs are provided to others to develop standalone applications. The project aims to provide online ID for each object, which is accessible via a link address. These objects can be linked with others via relationships and exchange information with other objects and humans. *Social web of thing* is a new platform developed by scientists at the Erosion Institute, which aims to use social media to greatly improve human presence. Interaction among objects is enabled using social media. Third-party applications like *Xively* and *Paraimpu* support the creation of web-based applications that can link smart objects in a social network. Pintus et al. [44] proposed *Paraimpu*, which added value to the smart objects using Http-enabled connection. The proposed architecture was based on Social Web of Thing concept. Using this platform, end users can register new smart objects and build a wide range of personalized applications in a user-friendly way. The main limitation of the proposed algorithm was that there is no mechanism for handling the heterogeneous devices and API offered for different architectures. However, using this approach, no social interaction and relationship can be utilized and hence restricts the user and device collaboration. Pintus et al. [45] proposed another improved architecture using the concept of Social Web of Thing. This architecture explains the way how different heterogeneous devices can be added to a social-enabled platform. It is a web-enabled platform that adds a virtual device, adds social value to the social device, shows how to collect information from the devices, and shares data from heterogeneous devices using programming panel.

Girau et al. [46] proposed an architecture called based on ThingSpeak server and the concept of SIoT. ThingSpeak server is used to manage the social interaction of the social objects and provide the features of how to add new devices and add rules about the social relationship. Web server provides the objects with the required information whenever this information is required by the smart object. However, the proposed platform does not define and determine the trustworthiness of the services received. Reference [47] also proposed a web-enabled platform for SIoT utilizing semantic web service and social network. The proposed architecture utilized the SN as a service creation platform where the end users/admin can create services for their smart devices. The main aims of this platform were to collect information from different devices and share them with friend's devices in a social network. Danielle Sheridan et al. [48] introduced the concept of using Twitter and IoT devices for the development of social IoT system. In this architecture, APIs provided by Twitter are utilized for human-to-machine communication. REST and MQTT protocols are used for human-to-machine communication, and the result indicates that the proposed system is ideal for SIoT networks. Byun et al. [34] proposed *Lilliput* for IoT devices. This architecture improves social graph by improving the social interaction among social objects. The proposed platform enables end users to develop SIoT-based application without prior knowledge of programming and skills. Three types of bi-directional relationships can be created, and they are divided into human to human, place to place, and object to object relationships. The main advantage of this platform is that it studies all aspects of social relationship and proposes an efficient hybrid model for both SN and IoT. The limitation of the proposed platform is that the temporal social relationship between people, devices, and locations is not expressed, and it can increase the cost because of the utilization of ML methods.

Zhang et al. [33] proposed an architecture design for SIoT application based on the concept of Social Web of Thing, Restful web services, and SN structure. The proposed platform makes use of web technologies and SN to bring the social relationship in smart objects. In this platform, semantic web technologies are used to convert the raw data into different natural languages allowing smart devices to interact with each other. The proposed platform also provides API to provide access to their database services for third-party users. *Socialite* is another platform proposed by Kim et al. [35] for the development of SIoT applications. In this platform, a set of relationship is predefined and can be used for the development of SIoT applications. It allows the user to integrate devices having different underlying architectures, with various types of interface, and to allow defining the relationships. *Socialite* attains effective RM for the SIoT by developing the relationship ontology. Girau et al. [49] introduced *Lysis* for the deployment of IoT applications. This is a cloud-based platform having four major features, i.e., social agent, PaaS model, reusability, and cloud storage for information storage and operation. Both users and developers use the PaaS for the development of applications. Relationships are established among social objects to locate information and make the network more scalable. Built-in templates are provided to the user and entire community to develop and deploy an SIoT-based application. Data generated by the devices are stored in the server, which is controlled by the developers. Cicirelli et al. [50] developed *iSapiens* (a java based) platform for implementing SIoT-based applications. iSapiens allow the user to add new objects having social capabilities and interaction with other devices and objects. The proposed platform allows the user to create smart environments and manage cloud storage and other resources. iSapiens is designed and implemented for the development of smart city services and applications [51]. Reference [52] proposed a platform in which users can automatically add new devices, allow end user to deploy different protocols, and analyze the result in a visualized form. In this system, MQTT and CoAP protocols can be implemented for data communication. The proposed platform is integrated with the social media application through which the user can get notifications about any event. The proposed framework was implemented for the smart home scenario. In this system, a cloud-based server (ThingSpeak) is used to collect the sensor data where these data are further analyzed (Table 9.3).

9.8 DATABASE OF SIoT

In literature, there are still very limited databases available for the analysis of SIoT networks, which are complete in all aspects as "Brightkite" [57–75]. As SIoT in almost at the development stage, different databases have been developed and proposed for the SIoT, which are available for the research community.

Marche et al. [6] created an SIoT database based on a real experiment in the city of Santander. Different objects are developed and deployed in the city, and the results are obtained. These results include object profile, their interaction, and types of relationships. These results are stored and used to generate databases that are publicly available for research communities. Similarly, Ref. [76] suggested an IoT dataset in which objects along with their social profiles are included in the dataset. Reference [76], combined the trust metric in it, applies different ML and DL techniques for authentication and validation. Table 9.4 shows a list of studied datasets of the SIoT dataset along with their brief description and features.

TABLE 9.3   SIoT Platform

Project/Company	Interaction	Interface	Social Relationship	Open Source	Application	Website
Xively	Minimal	–	No	Yes	Not specified	www.xively.com
Paraimpu [44]	Yes	Restful APi	No	Yes	Not specified	www.crs4.it/paraimpu
Toyota friend	Minimal	–	No	No	Yes	Twitter.com/toyotafriend
Social Internet of Things [46]	Yes	PaaS model	Yes	Yes	Yes	http://www.social-iot.org
Social robot	Yes	–	Not specified	Yes	No	Not specified
Everthng	Yes	–	Yes	Yes	Not specified	www.evrythng.com
Social Web of Things	Yes	–	Not specified	Yes	Not specified	Not specified
Social Home	Not specified	–	Not specified	No	Yes	Apps.facebook.com/mysocialhome
Social devices	Minimal	–	Only co-location relationship	No	Yes	Not specified
Socialite [35]	Yes	–	Yes	Yes	Yes	Not mentioned
Nike +	Yes	–	No	No	Yes	www.nikeplus.com
Lilliput [34]	Yes	Restful API	Yes	Yes	Not mentioned	Not mentioned
Social pal [38]	Yes	Restful API	Yes	Yes	Yes	Not mentioned
Kaa	Yes	Restful API	Yes	Yes	No	www.kaaproject.org

9.9 TRUST MANAGEMENT

Trust is a crucial element in today's era and technology, particularly in IoT systems and cloud computing [53], as it pertains to how objects interact with each other [54]. The lack of trust between objects in a socially interactive system can result in various important problems, e.g., loss of information privacy, security, safety, and unauthorized manipulation of information. Additionally, object owners may carry out detrimental actions such as self-promotion, bad-mouthing, and on-off attacks. Thus, evaluating trust-related issues among social objects is vital for SIoT to find the optimal interactions between customers and providers/developers. To create and develop a trustworthy connection, it is necessary to have a high level of confidence in the other social objects that are being connected. This can enhance the trustworthiness of communication by providing requests and separating malicious objects from trustworthy ones in the network [55].

Confidentiality is a key aspect of trust management (TM) in the IoT as it ensures that information is only accessible to authorized individuals at the appropriate time. This is because each object in an IoT system has its own vulnerabilities and may be susceptible to attacks. To prevent and block unauthorized access to data and network resources, it is necessary to implement a control system that enforces a security policy. This security policy should limit access to the network from potential attacks [56]. Furthermore, objects near each other often possess valuable information that they can share with other objects in a distributed, social manner in order to provide high-quality services. This exchange of information should be based on a careful selection of "friend" objects, as each object can discover relevant services by inquiring from its friends or "FoAF" (friends of a friend) to reduce the search area. Ultimately, it is crucial for each object to only exchange data and services with those objects that are trustworthy in order to establish a secure and reliable communication that can meet requested demands, thereby increasing the safety and security of SIoT networks.

9.9.1 Trust Types

In the context of the SIoT, there are different types of trust that are important to consider. Most of related articles, such as [54,55,57], mentioned trust as the main type:

1. **Transitive Trust:** This type of trust is established between different objects that may not know each other directly, e.g., if object A trusts B and B trusts C, then in this type of trust, A must be confident on C directly.

2. **Direct Trust:** Direct trust is based on direct mutual understanding and perception between smart objects.

3. **Indirect Trust:** Indirect trust is based on other smart object reputations and recommendations.

4. **Local Trust:** This type of trust is different for smart objects. For example, object A trusts B and object B trusts C, where C does not trust A.

5. **Subjective Trust:** This type of trust is based on individual object opinions, while objective trust between social objects is based on each other social/smart object's quality of service provided.

6. **Composite Trust:** This trust is a composite of friend recommendations and opinions, which may lead to establish trust or distrust about another object.

7. **Dynamic Trust:** This trust is not static during different time durations and may change over time, whether the conditions are altered.

Abdelghani et al. [58] proposed two types of trust between objects in the SIoT: Quality of Service (QoS) trust and social trust. QoS trust pertains to the ability of a device to provide high-quality service in response to a request for these services. It is evaluated using metrics such as reliability and cooperativeness. Social trust, on the contrary, is more common in the SIoT context and refers to the level of trust between the owners with each other. Trust is evaluated using factors such as connectivity and honesty. Reference [54] proposed to study the amount and degree of trust in the SIoT against different types of attacks such as On-Off selective forwarding attacks. This model is unique as it categorizes trust into three different types: centrality, energy, and service score. Centrality refers to the importance of smart/social objects for other objects to protect the SIoT network from malicious objects and reduce the number of communications between objects trying to achieve harmful goals to disturb the normal working of the network. Energy is another matric for trust evaluation, specifically in the case of On-Off attacks with in the network. Service score refers to the evaluation of the quality of service provided by an object. This dynamic trust model takes into account these three factors to provide a comprehensive and robust approach to trust evaluation in SIoT systems [59].

9.9.2 Trust Models

Evaluating trust models and building trust in the SIoT are closely related concepts. There are various models for establishing trust in SIoT, such as QoS trust, social trust, and dynamic trust models, and all these models need to be checked for accuracy and correction. Trust evaluation models are used to assess the trust models that are proposed for use in SIoT systems. These evaluations are necessary to ensure that the trust models used in the SIoT are reliable and accurate and that they can manage the trust between objects in the system. It is important to note that trust evaluation models are used to evaluate the trust models, not to establish trust between objects; they are two different things. Therefore, it is essential to use a robust and comprehensive evaluation process to ensure that the trust models used in SIoT systems can provide the necessary level of security, privacy, and reliability.

Many studies have proposed models for evaluating trust in SIoT, such as the one proposed by Nitti et al. [55]. They introduced a dynamic trust model. The model supports two types of trust evaluation: metric objective and subjective trust. The subjective model for the evaluation of trust has a much slower response time compared to the objective model, which processes and stores trust information in a DHT system that is visible to all network objects. The strength of this research is that it can separate distrustful objects from the network effects, by analyzing the technical trust metrics between related objects that cooperate. However, the weakness of this study is that it does not use direct observations to evaluate trust metrics and relies only on indirect trust observations. Reference [60] proposed a trust evaluation model for the SIoT that is based on objects' behavior. This model, called the SIoT Guarantor and Reputation Trust Evaluation Model, aims to provide a proper service response by evaluating the degree of trust between objects. The model employs techniques such as credit rating and reputation rating to assess trust. Objects that provide complete services are given more rank than those that do not collaborate or provide appropriate services. Objects with a lower rank are considered malicious. This approach is effective in identifying dishonest objects, but it does not consider all the important aspects of trust in large-scale networks.

Chen et al. [61] proposed a social trust evaluation model in SIoT networks that takes into account the interactions between owners of different objects. They identified three main factors that influence trust: Social Contact, Friendship, and Community of Interest. Friendship trust is based on the degree of closeness between the owners of two objects. Social Contact trust is based on the frequency of interactions between the owners. Community of Interest trust is based on the common interests shared by the owners. This model also considers the resiliency of the system against opportunistic service attacks. However, it does not consider different types of attacks and their potential impact on the trust evaluation. This may limit the effectiveness of the model in protecting against malicious objects or threats in the network. However, this study does not consider dynamic environmental factors, which may be a limitation. Sharma et al. [62] proposed a model that utilizes Fission Computing and edge-crowdsourcing network to evaluate trust and privacy. This model uses a combination of theoretical analysis and numerical simulations to evaluate the effectiveness of the proposed method using real data. Fission Computing helps balance the load in the network, while edge-crowdsourcing network allows for the evaluation of trust and privacy in a decentralized manner. This is the first model to use Fission Computing and edge-crowdsourcing network for trust and privacy evaluation in SIoT. The main advantage of this model is that it can effectively evaluate trust and privacy in a decentralized manner, which is essential for large-scale SIoT networks. Truong et al. [63] proposed a comprehensive approach for assessing trust in SIoT, which encompasses all components of the system. They developed a smart platform that evaluates trust services within the network, comprising three key elements: Reputation, Recommendation, and Knowledge. The Reputation aspect is based on user feedback, where the trustworthiness of an object is evaluated based on factors such as QoS, response time, and reliability. The Recommendation aspect is based on user recommendations to trust or distrust an object. The Knowledge aspect is based on the default knowledge of each object. The authors also used a car-sharing service use case to demonstrate the implementation and effectiveness of their approach. This approach is a holistic one, and it considers multiple factors and perspectives to evaluate trust, which makes it more reliable and accurate. Table 9.5 shows recent works related to the trust model.

TABLE 9.4 Available Dataset of SIoT

Serial No	Dataset Name	Company	Types of Features	Link Address
1	SIoT dataset	University of Cagliari	Complete database	www.social-iot.org/
2	SNAP	Bright light	Location based	www.snap.stanford.edu/data/loc-brighligth
3	MIT	Data mining on human mobility	Community based	http://Realitycommons.media.mit.edu
4	CRAWDAD	Cambridge/haggle	Object relationship	Crawded.org/Cambridge/haggle
5	CRAWDAD	Upb/hyccups	Social interaction	Crawded.org/keyword-socialnetwork.html
6	CRAWDAD	Cmu/hotspot	Location based	Crawded.org/cmu/hotspot/
7	SNAP	Facebook	Social circle	http://snap.stanford.edu/data/egonet-Facebook.html

TABLE 9.5 A Comparison between Trust Articles with Details

Article	Strength	Weakness	Main Topic	New Idea
[51]	High efficiency, adaptability, and reliability	No evaluation on the direct trust type	Trustworthiness management in the social IoT	New dynamic trust model
[60]	Model suitable to detect misbehaved objects	Not consider all main trust aspects	Based on guarantor and reputation-based trust mode	Guarantor and reputation-based trust model
[57]	High scalability and performance	Limited to opportunistic service attacks	Trust management and service composition for SOA-based IoT and its application	Not mentioned
[62]	Maximum availability Low complexity Minimum integration cost	Not mentioned	Edge-crowdsourcing in SIoT	Edge-crowdsourcing in SIoT
[63]	Improving trust evaluation High adaptability	Not any simulation tool and implement an approach	Trust service platform for SIoT	Trust service platform

9.10 WEB SERVICES

The main process for developing web services in recent technologies enables a system when a particular service is requested from the object until a smart object responds to it. The main procedure is further divided into three different sub-sections including service composition, selection, and composition. Composition is a process in which different social objects deliver the required services needed for the social object. This is done by comparing the characteristics of the objects with the requirements of the service request and selecting the object that best meets the requirements. Service selection finds the appropriate service matching the requested services. Service selection is a critical step in the web services process, as it ensures that the service provided is of high quality and meets the user's needs. Service composition is combining different services from different objects to offer an appropriate response to requested services. This is done by combining the capabilities of multiple objects to create a new service. Service composition ensures that the service provided has the desired quality of available service, functionalities, and feasibility. A single service cannot meet all objects' needs, and therefore, combining different services can provide a better solution to the user's request. In short, these all submodule plays important roles in the SIoT environment. This process is crucial as it ensures that the service provided is of high quality, meets user's needs, and provides better functionalities and feasibility. Different authors have proposed different types of service composition models, and here we discuss some recent developments in the direction of web services and services discovery (Figure 9.3).

Kouicem et al. [64] proposed a dynamic framework for service selection and composition in the SIoT based on a multi-agent dynamic structure that is implemented in the cloud computing paradigm. The framework uses large-scale methods to enhance system performance and improve QoS and other contextual metrics of service composition and selection. The multi-agent system is responsible for handling all the important operations of the network. To support these important operations, the authors implemented three different algorithms: plan generation, best selection, and updating QoS metrics after selection execution. The evaluation of their approaches was done in comparison to three similar approaches and studied the flexibility, scalability, and adaptability of their approaches. However, it should be noted that the framework proposed by Kouicem et al. [64] did not consider any models, and there is no exact statistics provided for the QoS evaluation metrics used in the study. Chen et al. [65] recommended a distributed structure for different operations based on object interconnections with each other based on three-dimensional structure and RESTful. Butt et al. [37] developed a context-aware protocol for service in SIoT, which aims to improve the main factors of service discovery such as reducing service

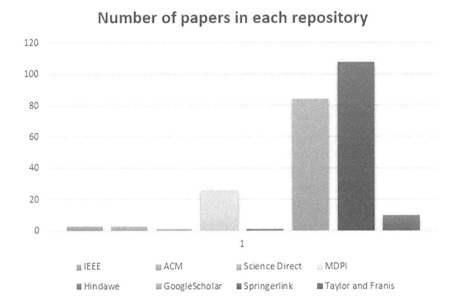

FIGURE 9.3   Web services process from request to response.

request delay, network stability, packet overhead and energy consumption. The proposed protocol, called Trendy, utilizes an efficient discovery technique along with a context-aware selection mechanism for IoT web services. This approach aims to increase the efficiency of service discovery by reducing the number of packets in the network, thus improving network scalability. Additionally, the protocol is designed to minimize the delay in service invocation and energy consumption during the network life cycle. Overall, this protocol aims to provide an efficient and effective solution for service discovery in IoT environments. Wei and Jin [66] proposed a concept of "contextual similarity" to measure the similarity between services and context information, which can be used to improve the accuracy and effectiveness of service discovery in IoT. They also proposed a context-aware service discovery algorithm that utilizes the contextual similarity concept, which can improve the matching between services and context information.

Furthermore, they evaluated the proposed context-aware service discovery algorithm using a simulated dataset and demonstrated that it can significantly improve the efficiency and effectiveness of service discovery in IoT. However, it is worth noting that this study does not provide any real-world evaluation or experimental results to validate the proposed approach. Xia et al. [67] proposed a service discovery mechanism called Scalable and Lightweight Service Discovery for SIoT that aims to improve the performance of service discovery in terms of three main metrics: success rate of queries, average path length of searches, and average number of relay nodes. The goal of SLSA is to provide an efficient and secure way of discovering web services in a large-scale network. The authors compared SLSA with three other mechanisms and evaluated their performance by classifying web services into 50 different subgroups, each containing 10 types of specific services. The results of the evaluation showed that SLSA has a higher success rate in queries, more energy effective in terms of the average number of relay nodes, and a shorter average path length of the discovery process compared to the other mechanisms. Additionally, the authors pointed out that SLSA's evaluation is based on a dynamic behavior in a large-scale network, and it can address resource limitations in SIoT. This study is significant as it provides a comprehensive evaluation of SLSA's performance and demonstrates its effectiveness in addressing the challenges of service discovery in SIoT. Reference [68] proposed a DNS-SD approach for discovering different web services for resource-constrained devices in SIoT environment. The proposed approach is based on combining three different protocols, known as *uBonjour*, to improve interoperability between different devices to discover desired services within the network. The main advantage of this approach is to promote the concept of service discovery process. However, they did not mention any possible flaws or limitations of the proposed approach in their study. Additionally, their implementation testbed is relatively small, and it would be beneficial to test the approach in a more extensive and realistic environment.

In addition, further investigation should be conducted to assess the scalability and robustness of the proposed approach in a large-scale IoT network. Reference [69] proposed an efficient method for sharing resources and selecting the appropriate objects to deliver a service based on QoS criteria in SIoT. They use a programming approach to develop their approach, and the strength of this study is its high level of flexibility in resource sharing.

However, they do not fully address the entire system in terms of QoS and leave it as a future area of research. The proposed solution aims to optimize the use of resources by considering the QoS criteria of the services and the devices, which leads to improved performance and better user satisfaction. Furthermore, the study highlights the importance of secure communication in resource sharing and selection in SIoT, which is a critical aspect to consider in the development of IoT systems. Reference [70] proposed an energy-efficient service composition method for SIoT services. This approach aims to minimize the number of different web services, which are very important to handle energy consumption for SIoT devices while satisfying all user requirements. A strength of this research is the high performance and energy efficiency as compared to other approaches; however, it does not take into account the security aspect of the service composition method. Reference [71] proposed a new algorithm for service composition that utilizes recommendation-based IoT services to meet the needs of users. They implemented a Typed Attribute Graph framework to recommend available services in IoT network, making it easier for smart objects to communicate with one another. A key aspect of their study is the use of an efficient algorithm to calculate service signatures, allowing for seamless service interactions. However, they did not perform any simulations to evaluate the effectiveness of their approach. It would be useful to have some form of simulation or testing to understand how well the proposed approach performs in a real-world scenario. Reference [72] proposed an approach using Genetic Programming along with a greedy search algorithm for web service composition. They tested their approach using built real test bed scenarios such as WSC-2008 and WSC-2009 to evaluate their performance. They compared their proposed approach with GP-based algorithm without greedy search implementation and found that the proposed method is more accurate and efficient as compared to that without a greedy search algorithm. The proposed algorithm also evaluated the approach in both dynamic and static environments.

9.11 ARCHITECTURE AND CLOUD-BASED SYSTEMS

In recent years, several architectures have been proposed to effectively implement the concept of Social Internet of Things (SIoT) across various domains. Some of these architectural models are discussed in the section below.

9.11.1 Computation Domain in SIoT

In the last decades, the development and significance of cloud computing have drawn the attention of researchers. Besides all these, the increasing number of connected devices and their access to connect to the high speed networks also increase the utilization and advantage of cloud computing [73]. Recently, cloud computing technologies have been widely applied due to high flexibility and cost-efficiency in which different computing and other computational resources work in a centralized manner.

To this, several different cloud-based computing paradigms have been proposed. In SIoT, there are a large number of heterogeneous devices sending their sensed data to the cloud for computation and different operations. The main goal of cloud computing is to provide a centralized location for computation which is accessed through internet [74].

Some of the other advantages of using cloud computing are being cost-effective and having high security, reliability, and high scalability [75]. Besides these advantages, cloud computing is not suitable for developing real-time applications because these storages are far away from the user's devices and have a WAN delay/latency [76,77].

This section reviews the latest cloud computation paradigm and Edge computing invested for the mentioned projects. In V. Doctor project, Cloud-based solution is presented in which unlimited requests from patients are handled.

In Ref. [78], scalability is improved by using cloud as there are a large number of simultaneous search queries from different devices. In Ref. [79], the author developed a smart home based on SIoT to optimize the energy usage of the building. There are other methods for energy management but in this chapter, the author developed a user-centric approach while keeping the comfort level on first priority. In this project, a social virtual object is created, which represents a real-world object in the cloud. Different sensors are deployed, which send data to the cloud-based server (Lysis Platform).

In social beach application [80], the concept of SIoT was used to find the best-ever beach according to user preference. This system consists of a control unit along with ultraviolet sensors, sensor for humidity detection, mobile devices, beach station, and camera device for crowd estimation using a combination of support vector machine and computer vision algorithms. First, data are collected from these devices and then send the data to the cloud-based server (Lysis Platform) where information about the beach and user preference is estimated, and the users are informed about the nearest beach according to their choice. In smart office project [81], social objects are organized by their location and their services. Their services are treated as their operations. In smart office, they used the Xively Platform 1 as the cloud server, which is a real-time network. In that platform, data can be exchanged using the standard REST protocol. In Airport Dynamic Social [82], a cloud-based system was developed, which enables the end user to interact with the smart objects in the airport, i.e., sensors installed at check-in for airport counters, boarding gates, flights information, and other services within the airport. Additionally, the developed application aims at aiding from Internet Protocol version 6 in order to show the effectiveness of the proposed system by adding several heterogeneous devices in a smart space. In Ref. [50], the author proposed a smart home solution based on the concept of SIoT and Edge computing. Edge computing [83] is a powerful and dynamic technology offering a high speed computation and storage facilities at the edge for developing real-time applications. In this application [50], a smart environment is created as a proof of concept, specifically for managing and recovery in a dangerous situations. In MagicHome application [33], the author proposed a cloud-based approach for developing a smart home application based on the concept of SIoT. In this application, different sensors are deployed at different locations in the home which continually sense environmental parameters and send it to the centralized domain. The author in Ref. [84] developed two different applications, namely, Vehicle Diagnosis System and Smart Parking based on the concept of SIoT. In these applications, data are collected to a cloud-based server for car diagnosis services and vacant parking location information. The application enables the owner to monitor the Engine Control Unit parameters and can be viewed via a webpage available from a personal computer and other networking devices or even other rendering devices in the car. Similarly, in Smart Parking application, information is provided to the users about vacant locations. In these applications, the data are first sent to cloud-based platform (Lysis) where these data are processed.

9.11.2 Application Domains of SIoT

There are several application scenarios where heterogonous devices are involved in building social relationships among them. Several types of relationships can be established according to the nature and context of interactions [85]. In this section, some of the use cases of implementing socialized concepts in IoT are discussed. Healthcare domain is one of the most important aspects of human life.

Nowadays, SIoT has gained much attraction in healthcare domain [86]. In healthcare domain, it is necessary to sense and receive information about elder persons' health whenever they suffer from some diseases. Some of the solutions based on SIoT are developed. In V.Doctor solution, a framework is introduced to integrate the concept of SIoT and E-health for monitoring elderly persons having some diseases [87]. This system keeps monitoring the object's health and provides medical guidelines and can discover the right persons who can help them in critical situations. In healthcare domain, another SIoT-based application called Elder Monitoring System based on SIoT has been developed. In this application, physical and environmental data are collected and sent to the right person [88]. The proposed system has been designed in such a way that new social services and social devices can be easily integrated and removed from the networks. In this system, heterogeneous devices having different architectures and protocols can directly share data that fulfill the SIoT requirements. Similarly, the physical layered aided projects in which social networks are employed as a trustworthy platform for sharing the data of patients to the healthcare providers. As patients' data are highly private and confidential to the patients, the author [89] introduced security techniques for protecting this secure information in the SIoT system. This paper introduces two different solutions for healthcare that classify heart disease and the second for brain tumor classification and identification. The result shows the effectiveness of the SIoT application in healthcare domain. In Ref. [90], the author proposed a system called FriendCare-ALL for elder person monitoring. In this application, socially enabled IoT devices are created and deployed in a smart environment, especially in smart home scenario applications. Human activities of elder persons are monitored, and whenever any abnormal activities occur, the developed system generates an alarm. To improve the detection of abnormal activities, two different ML algorithms were also used to correctly identify the fall activity of the elder person.

Intelligent transportation is another important domain in which IoT plays an important role [91]. Number of vehicles are increasing every day, which leads to an increase in traffic, accident, and air pollution. There IoT can be efficiently employed to solve these mentioned issues. In this regard, the concept of IoV has evolved [92], and its different social variants have been investigated in Ref. [93]. Indeed, SIoT is introduced in the ITS domain to improve traffic management and road safety systems. To improve the existence of ITS, research in the field of how to integrate the concept of SIoT in the field of ITS and different research projects are created to improve the ITS systems. SIoT is applied in the

Vehicular Social Network project for the interaction of vehicles and raising the level of driving knowledge [94]. Similarly, in tNote project [36], an SIoV based on Vehicular Ad Hoc Networks has been proposed. In this system, users (vehicle) share their information with other vehicles. In Ref. [84], the author developed two different applications based on SIoT in the SIoV domain. The first application is a Vehicle Diagnostic application, which allows vehicle owners to access information about their vehicles through a proposed SIoV platform. This application enables users to monitor various Engine Control Unit parameters, such as engine revolution speed, vehicle speed, and oil temperature, through an onboard unit. This information is then made available to other social objects via a web page, a smartphone application, or even the vehicle's own display. The second application is a Smart Parking application, which demonstrates how SIoV applications can be used to share important information about car parking spaces in a smart city. In this application, social relationships are created among social vehicles and smart parking locations to share alerts about the availability of free parking spaces. Only registered vehicles can access information about free parking locations by creating a social relationship, allowing users to make informed decisions about where to park at a given time. SIoT can improve the quality of life by providing smart services and applications such as these.

Smart Place is a concept where the use of IoT devices and advanced applications is applied to improve the life standard [95]. Smart building has various capabilities [94] as compared to other traditional buildings including safety, temperature control, etc. Several projects employed the concept of SIoT in a smart place to enhance the existing system. One solution based on SIoT was proposed by Ref. [51], which aims to study the security issue, large-scale heterogeneous device management, friendship discovery, and trust in large-scale smart environment. In this SIoT-based project, a new middleware layered is introduced to efficiently manage the heterogeneous devices and mobility model of the devices in any smart environments. In Ref. [79], the author efficiently applied the concept of SIoT paradigm for a Smart Heating, Ventilation, and Air Conditioning system, which is used to develop a tradeoff model for energy being consumed, and the thermal comfort of users inside a building is achieved. In Ref. [79], the author first developed the thermal profile of the considered building which is being characterized and also other parameters such as external environment and room occupancy. From the simulation results, the implemented system can reduce energy costs as compared to other traditional methods while keeping the user's comfort levels and needs. In order to ease interactions and context awareness among different sensors and actuators, the concept of SIoT is applied. In Ref. [96], the author proposed an SIoT platform for the detection and identification of vacant space in smart car parking. In this system, a platform called Lysis [49], which is a cloud-based platform, is used to create a virtual object in the cloud, which is involved in the real world for smart parking. Magnetometer was used to detect the vacant space in the car parking, and a control dashboard was designed to manage the vacant car space in the smart car parking. Experimental testbed along with an android application was also implemented to prove the effectiveness of the proposed system. In Ref. [97], the author proposed an SIoT platform for the monitoring of the sea and providing meaningful information to the visitor. In this system, a small buoy was created to be placed on different positions at the sea which are integrated into a cloud-based platform for managing and information to the visitors. Another SIoT-based beach monitoring system was proposed by Ref. [80]. In this system, the concept of SIoT was used to find the best beach according to user preference. This system consists of a control unit along with ultraviolet sensors, sensor for humidity detection, mobile devices, beach station, and camera device for crowd estimation using a combination of support vector machine and computer vision algorithms. Also in this system, local environmental data are collected using these sensors. Data from those devices are sent to a cloud-based platform that provides users about the beach and suggestions about where to go based on the preferences set by the users. These preferences are weather information, total time taken to reach a beach, crowdedness information, and so on.

In Ref. [81] the author developed an SIoT-based smart office environment in which different social objects are categorized based on their dynamic location. In this application, social objects are organized by their location and their services. Their services are treated as their operations. In this paper, the author has used the Xively Platform 2 as the cloud server, which is a real-time network. In that platform, data can be exchanged using the standard REST protocol. In Ref. [82], an application is provided for smart airport, namely, Airport Dynamic Social. The aim of the developed system is to allow users to directly interact with all available social objects and their smart services in smart airport. These social objects and smart services are different sensors at different sections within the airport. Additionally, this application takes benefits from IPV6 in order to cope with several heterogeneous devices in a smart space as well as to manage resource-constraint devices. In Ref.[33], the author developed and deployed an SIoT-based smart home solution called MagicHome. This application was based on the Social Web of Thing Framework, which relies on the Social Network and Restful Web Service. In MagicHome, there are lot number of sensors deployed for sensing purposes. In this application, a social agent posts their sensed information to other social things who are friends, follow new feeds, and chat with other social objects. In short, MagicHome is a society whose members are the different social devices having their own social networks? In Ref. [50], the author developed a project based on SIoT concept called iSapiens. ISapiens implements the concept of SIoT that allows objects to establish social relationships to handle emergency (Table 9.6).

9.12 SIoT CHALLENGES AND OPEN ISSUE

There are billions of smart devices, which are expected to be connected to the internet having different architecture and nature. In the SIoT environment, social objects can create a relationship with other social objects having different mobility models and nature. At present, the SIoT is still very immature and is at a very early stage of development. Due to the dynamic nature of SIoT, there exist some challenges that should be considered while developing an SIoT application. Some of these challenges are as follows.

9.12.1 Heterogeneous Devices

In the SIoT environment, there are many devices like sensors, actuators, computer system, personal devices, RFID devices, and other connecting devices that communicate with each other regardless of their underlying environments and their operating platforms. In real

TABLE 9.6   Application Field of SIoT in Different Areas

Project	Domain	Publisher	References	Computation Domain
V.Doctor	Healthcare	IEEE	[78]	Cloud based
Elder monitoring system	Healthcare	IEEE	[88]	Not mentioned
PHY Aided	Healthcare	IEEE	[89]	Not mentioned
Social Home	Smart place	IEEE	[79]	Cloud based
Smart beach	Smart place	IEEE	[80]	Cloud based
Smart office	Smart place	IEEE	[81]	Cloud based
Social airport	Smart place	Elsevier	[82]	Cloud based
iSapiens	Smart place	IEEE	[51]	Edge based
MagicHome	Smart place	ACM	[33]	Not mentioned
TNote	Intelligent Traffic	Elsevier	[84]	Cloud based
Social IoV	Intelligent traffic	Elsevier	[37]	Cloud based
SIoT based smart parking	Intelligent car parking	Elsevier	[96]	Cloud based
Sea monitoring system	Monitoring system	IEEE	[97]	Cloud based

time, the interoperability of these heterogenous devices is a challenging task. The data transferred from each device have their own formats according to their underlying architecture and operating system, and it is very difficult in such an environment to manage, analyze, and process these data for decision-making.

In the SIoT environment, the total time taken by the social objects to communicate with each other and to understand each other in such heterogenous environments also degrades the overall performance of the system. In the SIoT environment, different smart objects and humans as an entity should interact in a machine and human-friendly manner. In fact, there is no total interoperability for diverse devices architectures of different mechanisms that exist for IoT which effect the development of SIoT applications [18,98].

9.12.2 System for Dynamic Friendship Management

Atzori et al. [12] suggested that social objects can create social relationship with other objects in a dynamic way for service discovery. However, the proposed system does not support dynamic relationship in such a dynamic way. Most of the objects require dynamic object selection for information retrieval, and objects need to have the ability to assume new with other social objects. Reference [99] used semantic web ontology to establish a social relationship for efficient service acquisition and provisioning. Other such social relationship types should be established to achieve multicast features and classified results. There is a need for some strategies that focus on the development of relationship for the specific task, gaining better services and selecting a friend that improves the network navigation.

9.12.3 Optimal Link Selection Strategy

Dynamic discovery of services in the SIoT network is one of the most critical issues. This has arisen due to the fact of Friend of a Friend approach for the service discovery. However, this greatly increases the search time because each object searches all the friends of other social objects. Therefore, finding a solution for the link selection problem in the SIoT needs to be addressed. For this, Ref. [100] proposed an enhanced link selection algorithm based on the application of a genetic algorithm [101].

9.12.4 Data Handling and Data Management

In SIoT, several heterogonous devices (social objects) continuously send their sensed data to other social objects in their own formats. In real time, it is very difficult to filter these data, to process these data, and to produce output for the end users. Semantics of the data should maintain reliability, validity, integrity, consistency, and sensitivity [18]. There should be a dynamic mechanism to handle these dynamic data and to manage these data in such a way that it increases efficiency and reduces redundancy among social devices. In the SIoT environment, when the social object performs multiple tasks at a single time, the SIoT system should remain effective in an efficient way [18].

9.12.5 Energy Consumption Management

In the SIoT environment, when a social object is deployed, it is very difficult to change the batteries of these devices without degrading the performance of the whole system, especially for different sensors that are already deployed in the field [98]. There are several challenges while deploying sensors as they need maintenance time by time.

9.12.6 Trust Management, Security, and Privacy

As in the SIoT environment, there are different social nodes that make relationships with other social nodes in friendly ways. This interaction in a social way poses different security and privacy issues. In such scenarios, many different questions arise such that:

- What will happen to the collected data if it is not used by other social objects?

- What degree of privacy is maintained while creating social relationship with other social nodes?

- How trustworthy the collected data are?

In the SIoT environment, issues related to privacy and trust arise as there are specific rules to which the social object should interact with other social objects because of security concerns that arise when data are freely accessed between devices.

9.12.7 Network Navigability

In SIoT applications, social objects offer their services, and these services are required of other social objects who are friends using their friendship circle in a distributed way. As in SIoT, there are large number of social objects connected to each other, and every object has a large number of friends making it difficult for other social objects to search a specific service. This process slow down the network performance [102]. Network navigability is an important research issue in SIoT that should be considered while developing SIoT-based applications. Some of the research works were done in Ref. [102] in SIoT for improving the network navigation which can serve as a reference model for network navigation in SIoT, but still there is a need for an efficient navigational scheme.

9.13 CONCLUSION

In few last decades, the number of smart objects connected to the internet grows exponentially. Social Internet of Things is a new paradigm by which different smart objects have become social objects and created social relationship with each other and with their owner. This chapter has reviewed the latest papers regarding the development of Social Internet of Things, types of relationships, different platforms and SIoT architecture, and application of SIoT in different domains. Finally, we have discussed several challenges and issues regarding SIoT.

REFERENCES

[1] K. Ashton et al., "That Internet of Things," *RFID Journal*, vol. 22, no. 7, pp. 97–114, 2009.

[2] S. Efremov, N. Pilipenko, and L. Voskov, "An integrated approach to common problems in the Internet of Things," *Procedia Engineering*, vol. 100, pp. 1215–1223, 2015.

[3] U. Cisco, "Cisco annual internet report (2018-2023) white paper. 2020," *Acessado em*, vol. 10, no. 01, pp. 1–35, 2021.

[4] G. I. Top, "Strategic IoT technologies and trends," in *Analysts Explore Internet of Things Opportunities and Pitfalls at Gartner Symposium/ITxpo* 2018, November 4–8, 2018, Barcelona, p. 10.

[5] J. Gubbi, R. Buyya, S. Marusic, and M. Palaniswami, "Internet of Things (IoT): A vision, architectural elements, and future directions," *Future Generation Computer Systems*, vol. 29, no. 7, pp. 1645–1660, 2013.

[6] D. Goad and U. Gal, "IoT design challenges and the social IoT solution," Twenty-third Americas Conference on Information Systems, Boston, AIS Electronic Library (AISeL), 2017.

[7] G. Panda, B. Tripathy, and M. Padhi, "Evolution of social IoT world: Security issues and research challenges," *Internet of Things (IoT)*, pp. 77–98, 2017.

[8] L. Atzori, A. Iera, and G. Morabito, "The Internet of Things: A survey," *Computer Networks*, vol. 54, no. 15, pp. 2787–2805, 2010.

[9] D. Gupta et al., "Energy saving implementation in hydraulic press using industrial Internet of Things (IIoT)," *Electronics*, vol. 11, no. 23, p. 4061, 2022.

[10] T. Mazhar et al., "Analysis of challenges and solutions of IoT in smart grids using AI and machine learning techniques: A review," *Electronics*, vol. 12, no. 1, p. 242, 2023.

[11] R. Pal, D. Adhikari, M. B. B. Heyat, I. Ullah, and Z. You, "Yoga meets intelligent Internet of Things: Recent challenges and future directions," *Bioengineering*, vol. 10, no. 4, p. 459, 2023.

[12] L. Atzori, A. Iera, G. Morabito, and M. Nitti, "The social Internet of Things (SIoT)-when social networks meet the Internet of Things: Concept, architecture and network characterization," *Computer Networks*, vol. 56, no. 16, pp. 3594–3608, 2012.

[13] S. Rho and Y. Chen, "Social Internet of Things: Applications, architectures and protocols," *Future Generation Computer Systems*, vol. 82, pp. 667–668, 2018.

[14] P. Kumaran and R. Sridhar, "Social Internet of Things (SIoT): Techniques, applications and challenges," in *2020 4th International Conference on Trends in Electronics and Informatics (ICOEI) (48184)*, Tirunelveli, India, 2020, pp. 445–450.

[15] A. P. Fiske, "The four elementary forms of sociality: Framework for a unified theory of social relations," *Psychological Review*, vol. 99, no. 4, p. 689, 1992.

[16] B. Tripathy, D. Dutta, and C. Tazivazvino, "On the research and development of social Internet of Things," in Constandinos X. Mavromoustakis, George Mastorakis, and Jordi Mongay Batalla (eds.). *Internet of Things (IoT) in 5G Mobile Technologies*, Springer, Switzerland, 2016, pp. 153–173.

[17] J. Formo, J. Laaksolahti, and M. Gårdman, "Internet of Things marries social media," in *Proceedings of the 13th International Conference on Human Computer Interaction with Mobile Devices and Services*, Stockholm, Sweden, 2011, pp. 753–755.

[18] A. M. Ortiz, D. Hussein, S. Park, S. N. Han, and N. Crespi, "The cluster between Internet of Things and social networks: Review and research challenges," *IEEE Internet of Things Journal*, vol. 1, no. 3, pp. 206–215, 2014.

[19] N. Gulati and P. D. Kaur, "When things become friends: A semantic perspective on the Social Internet of Things," in Bijaya Ketan Panigrahi, Munesh C. Trivedi, Krishn K. Mishra, Shailesh Tiwari, and Pradeep Kumar Singh (eds.). *Smart Innovations in Communication and Computational Sciences*, Springer, Singapore, 2019, pp. 149–159.

[20] T. Mazhar et al., "Analysis of IoT security challenges and its solutions using artificial intelligence," *Brain Sciences*, vol. 13, no. 4, p. 683, 2023.

[21] T.-Y. Chung et al., "MUL-SWoT: A social web of things platform for internet of things application development," in *2014 IEEE International Conference on Internet of Things (iThings), and IEEE Green Computing and Communications (GreenCom) and IEEE Cyber, Physical and Social Computing (CPSCom)*, Taipei, Taiwan, 2014, pp. 296–299.

[22] I. Mashal, O. Alsaryrah, T.-Y. Chung, C.-Z. Yang, W.-H. Kuo, and D. P. Agrawal, "Choices for interaction with things on Internet and underlying issues," *Ad Hoc Networks*, vol. 28, pp. 68–90, 2015.

[23] C. Perera, A. Zaslavsky, P. Christen, and D. Georgakopoulos, "Context aware computing for the Internet of Things: A survey," *IEEE Communications Surveys & Tutorials*, vol. 16, no. 1, pp. 414–454, 2013.

[24] J. Ye, S. Dobson, and S. McKeever, "Situation identification techniques in pervasive computing: A review," *Pervasive and Mobile Computing*, vol. 8, no. 1, pp. 36–66, 2012.

[25] Y. LeCun, Y. Bengio, and G. Hinton, "Deep learning," *Nature*, vol. 521, no. 7553, pp. 436–444, 2015.

[26] M. Kranz, P. Holleis, and A. Schmidt, "Embedded interaction: Interacting with the Internet of Things," *IEEE Internet Computing*, vol. 14, no. 2, pp. 46–53, 2009.

[27] J. Kleinberg, "The convergence of social and technological networks," *Communications of the ACM*, vol. 51, no. 11, pp. 66–72, 2008.

[28] J. An, X. Gui, W. Zhang, and J. Jiang, "Nodes social relations cognition for mobility-aware in the internet of things," in *Internet of Things (iThings/CPSCom), 2011 International Conference on and 4th International Conference on Cyber, Physical and Social Computing*, Dalian, China, 2011, pp. 687–691.

[29] W. Z. Khan, M. Y. Aalsalem, M. K. Khan, and Q. Arshad, "When social objects collaborate: Concepts, processing elements, attacks and challenges," *Computers & Electrical Engineering*, vol. 58, pp. 397–411, 2017.

[30] D. Dutta, T. Ch, S. Das, and B. Tripathy, "Social Internet of Things (SIoT): Transforming smart object to social object," in *NCMAC Conference*, Washington, USA, 2015.

[31] J. Guth et al., "A detailed analysis of IoT platform architectures: Concepts, similarities, and differences," in Beniamino Di Martino, Kuan-Ching Li, Laurence T. Yang, and Antonio Esposito (eds.). *Internet of Everything*, Springer, Singapore, 2018, pp. 81–101.

[32] L. Atzori, A. Iera, and G. Morabito, "Siot: Giving a social structure to the Internet of Things," *IEEE Communications Letters*, vol. 15, no. 11, pp. 1193–1195, 2011.

[33] C. Zhang, C. Cheng, and Y. Ji, "Architecture design for social web of things," in *Proceedings of the 1st International Workshop on Context Discovery and Data Mining*, Beijing, China, 2012, pp. 1–7.

[34] J. Byun, S. H. Kim, and D. Kim, "Lilliput: Ontology-based platform for IoT social networks," in *2014 IEEE International Conference on Services Computing*, Anchorage, AK, USA, 2014, pp. 139–146.

[35] J. E. Kim, A. Maron, and D. Mosse, "Socialite: A flexible framework for social Internet of Things," in *2015 16th IEEE International Conference on Mobile Data Management*, Pittsburgh, PA, USA, 2015, pp. 94–103.

[36] K. M. Alam, M. Saini, and A. El Saddik, "tnote: A social network of vehicles under Internet of Things," in *International Conference on Internet of Vehicles*, Beijing, China, 2014, pp. 227–236.

[37] T. A. Butt, R. Iqbal, S. C. Shah, and T. Umar, "Social internet of vehicles: Architecture and enabling technologies," *Computers & Electrical Engineering*, vol. 69, pp. 68–84, 2018.

[38] F. Amin and G. S. Choi, "Social pal: A combined platform for Internet of Things and social networks," in *2020 5th International Conference on Computer and Communication Systems (ICCCS)*, Shanghai, China, 2020, pp. 786–790.

[39] O. Voutyras, P. Bourelos, S. Gogouvitis, D. Kyriazis, and T. Varvarigou, "Social monitoring and social analysis in Internet of Things virtual networks," in *2015 18th International Conference on Intelligence in Next Generation Networks*, Paris, France, 2015, pp. 244–251.

[40] E. A. Kosmatos, N. D. Tselikas, A. C. Boucouvalas, "Integrating RFIDs and smart objects into a UnifiedInternet of Things architecture," *Advances in Internet of Things*, vol. 2011, pp. 5–12, 2011.

[41] O. Voutyras, P. Bourelos, D. Kyriazis, and T. Varvarigou, "An architecture supporting knowledge flow in Social Internet of Things systems," in *2014 IEEE 10th International Conference on Wireless and Mobile Computing, Networking and Communications (WiMob)*, Larnaca, Cyprus, 2014, pp. 100–105.

[42] M. Roopa, S. Pattar, R. Buyya, K. R. Venugopal, S. Iyengar, and L. Patnaik, "Social Internet of Things (SIoT): Foundations, thrust areas, systematic review and future directions," *Computer Communications*, vol. 139, pp. 32–57, 2019.

[43] R. Abdul, A. Paul, J. Gul M, W.-H. Hong, and H. Seo, "Exploiting small world problems in a SIoT environment," *Energies*, vol. 11, no. 8, p. 2089, 2018.

[44] A. Pintus, D. Carboni, and A. Piras, "Paraimpu: A platform for a social web of things," in *Proceedings of the 21st International Conference on World Wide Web*, Lyon, France, 2012, pp. 401–404.

[45] A. Piras, D. Carboni, and A. Pintus, "A platform to collect, manage and share heterogeneous sensor data," in *2012 Ninth International Conference on Networked Sensing (INSS)*, Antwerp, Belgium, 2012, pp. 1–2.

[46] R. Girau, M. Nitti, and L. Atzori, "Implementation of an experimental platform for the social Internet of Things," in *2013 Seventh International Conference on Innovative Mobile and Internet Services in Ubiquitous Computing*, Taichung, Taiwan, 2013, pp. 500–505.

[47] V. Beltran, A. M. Ortiz, D. Hussein, and N. Crespi, "A semantic service creation platform for Social IoT," in *2014 IEEE World Forum on Internet of Things (WF-IoT)*, Seoul, Korea (South), 2014, pp. 283–286.

[48] D. Sheridan, A. A. Simiscuka, and G.-M. Muntean, "Design, implementation and analysis of a Twitter-based social IoT network," in *2019 International Conference on Sensing and Instrumentation in IoT Era (ISSI)*, Lisbon, Portugal, 2019, pp. 1–6.

[49] R. Girau, S. Martis, and L. Atzori, "Lysis: A platform for IoT distributed applications over socially connected objects," *IEEE Internet of Things Journal*, vol. 4, no. 1, pp. 40–51, 2016.

[50] F. Cicirelli, A. Guerrieri, G. Spezzano, A. Vinci, O. Briante, and G. Ruggeri, "iSapiens: A platform for social and pervasive smart environments," in *2016 IEEE 3rd World Forum on Internet of Things (WF-IoT)*, 2016, Reston, VA, USA, pp. 365–370.

[51] F. Cicirelli, A. Guerrieri, G. Spezzano, and A. Vinci, "An edge-based platform for dynamic smart city applications," *Future Generation Computer Systems*, vol. 76, pp. 106–118, 2017.

[52] B.-S. Chen, V. A. Kshirsagar, and S.-C. Lo, "Platform design for social Internet of Things," in *2017 IEEE International Conference on Consumer Electronics-Taiwan* (*ICCE-TW*), Taipei, Taiwan, 2017, pp. 67–68.

[53] V. Mohammadi, A. M. Rahmani, A. M. Darwesh, and A. Sahafi, "Trust-based recommendation systems in Internet of Things: A systematic literature review," *Human-Centric Computing and Information Sciences*, vol. 9, no. 1, pp. 1–61, 2019.

[54] A. M. Kowshalya and M. Valarmathi, "Trust management in the social Internet of Things," *Wireless Personal Communications*, vol. 96, no. 2, pp. 2681–2691, 2017.

[55] M. Nitti, R. Girau, and L. Atzori, "Trustworthiness management in the social Internet of Things," *IEEE Transactions on Knowledge and Data Engineering*, vol. 26, no. 5, pp. 1253–1266, 2013.

[56] A. Kurniawan and M. Kyas, "A trust model-based Bayesian decision theory in large scale Internet of Things," in *2015 IEEE Tenth International Conference on Intelligent Sensors, Sensor Networks and Information Processing* (*ISSNIP*), Singapore, 2015, pp. 1–5.

[57] M. Nitti, "Managing the Internet of Things based on its social structure," 2014.

[58] W. Abdelghani, C. A. Zayani, I. Amous, and F. Sèdes, "Trust management in social Internet of Things: A survey," in *Conference on e-Business, e-Services and e-Society*, Swansea, UK, 2016, pp. 430–441.

[59] A. Meena Kowshalya and M. Valarmathi, "Trust management for reliable decision making among social objects in the Social Internet of Things," *IET Networks*, vol. 6, no. 4, pp. 75–80, 2017.

[60] H. Xiao, N. Sidhu, and B. Christianson, "Guarantor and reputation based trust model for social Internet of Things," in *2015 International Wireless Communications and Mobile Computing Conference* (*IWCMC*), 2015, pp. 600–605.

[61] R. Chen, J. Guo, and F. Bao, "Trust management for SOA-based IoT and its application to service composition," *IEEE Transactions on Services Computing*, vol. 9, no. 3, pp. 482–495, 2014.

[62] V. Sharma, I. You, D. N. K. Jayakody, and M. Atiquzzaman, "Cooperative trust relaying and privacy preservation via edge-crowdsourcing in social Internet of Things," *Future Generation Computer Systems*, vol. 92, pp. 758–776, 2019.

[63] N. B. Truong, T.-W. Um, and G. M. Lee, "A reputation and knowledge based trust service platform for trustworthy social Internet of Things," in *Innovations in Clouds, Internet and Networks* (*ICIN*), 2016, Paris, France, pp. 104–111.

[64] A. Kouicem, A. Chibani, A. Tari, Y. Amirat, and Z. Tari, "Dynamic services selection approach for the composition of complex services in the web of objects," in *2014 IEEE World Forum on Internet of Things* (*WF-IoT*), Seoul, Korea (South), 2014, pp. 298–303.

[65] G. Chen, J. Huang, B. Cheng, and J. Chen, "A social network based approach for IoT device management and service composition," in *2015 IEEE World Congress on Services*, New York, NY, USA, 2015, pp. 1–8.

[66] Q. Wei and Z. Jin, "Service discovery for Internet of Things: A context-awareness perspective," in *Proceedings of the Fourth Asia-Pacific Symposium on Internetware*, New York, NY, USA, 2012, pp. 1–6.

[67] H. Xia, C. Hu, F. Xiao, X. Cheng, and Z. Pan, "An efficient social-like semantic-aware service discovery mechanism for large-scale Internet of Things," *Computer Networks*, vol. 152, pp. 210–220, 2019.

[68] R. Klauck and M. Kirsche, "Bonjour contiki: A case study of a DNS-based discovery service for the Internet of Things," in *International Conference on Ad-Hoc Networks and Wireless*, Belgrade, Serbia, 2012, pp. 316–329.

[69] Z. U. Shamszaman and M. I. Ali, "Toward a smart society through semantic virtual-object enabled real-time management framework in the social Internet of Things," *IEEE Internet of Things Journal*, vol. 5, no. 4, pp. 2572–2579, 2017.

[70] T. Baker, M. Asim, H. Tawfik, B. Aldawsari, and R. Buyya, "An energy-aware service composition algorithm for multiple cloud-based IoT applications," *Journal of Network and Computer Applications*, vol. 89, pp. 96–108, 2017.

[71] M. Le Pallec, M. O. Mazouz, and L. Noirie, "Physical-interface-based IoT service characterization," in *Proceedings of the 6th International Conference on the Internet of Things*, New York, NY, USA, 2016, pp. 63–71.

[72] H. Ma, A. Wang, and M. Zhang, "A hybrid approach using genetic programming and greedy search for QoS-aware web service composition," in Abdelkader Hameurlain, Josef Küng, Roland Wagner, Hendrik Decker, Lenka Lhotska, and Sebastian Link (eds.). *Transactions on Large-Scale Data-and Knowledge-Centered Systems XVIII*, Springer, Berlin, Germany , 2015, pp. 180–205.

[73] H. Vahdat-Nejad, Z. Mazhar-Farimani, and A. Tavakolifar, "Social Internet of Things and new generation computing-A survey," in Aboul Ella Hassanien, Roheet Bhatnagar, Nour Eldeen M. Khalifa, and Mohamed Hamed N. Taha (eds.). *Toward Social Internet of Things (SIoT): Enabling Technologies, Architectures and Applications*, Springer, Switzerland, 2020, pp. 139–149.

[74] B. P. Rimal, E. Choi, and I. Lumb, "A taxonomy and survey of cloud computing systems," in *2009 Fifth International Joint Conference on INC, IMS and IDC*, Seoul, Korea (South), 2009, pp. 44–51.

[75] C. Wang, Q. Wang, K. Ren, N. Cao, and W. Lou, "Toward secure and dependable storage services in cloud computing," *IEEE Transactions on Services Computing*, vol. 5, no. 2, pp. 220–232, 2011.

[76] W. Yu et al., "A survey on the edge computing for the Internet of Things," *IEEE Access*, vol. 6, pp. 6900–6919, 2017.

[77] Y. Mao, C. You, J. Zhang, K. Huang, and K. B. Letaief, "A survey on mobile edge computing: The communication perspective," *IEEE Communications Surveys & Tutorials*, vol. 19, no. 4, pp. 2322–2358, 2017.

[78] G. Ruggeri and O. Briante, "A framework for iot and e-health systems integration based on the social Internet of Things paradigm," in *2017 International Symposium on Wireless Communication Systems (ISWCS)*, Bologna, Italy, 2017, pp. 426–431.

[79] C. Marche, M. Nitti, and V. Pilloni, "Energy efficiency in smart building: A comfort aware approach based on Social Internet of Things," in *2017 Global Internet of Things Summit (GIoTS)*, Geneva, Switzerland, 2017, pp. 1–6.

[80] R. Girau, E. Ferrara, M. Pintor, M. Sole, and D. Giusto, "Be right beach: A social IoT system for sustainable tourism based on beach overcrowding avoidance," in *2018 IEEE International Conference on Internet of Things (iThings) and IEEE Green Computing and Communications (GreenCom) and IEEE Cyber, Physical and Social Computing (CPSCom) and IEEE Smart Data (SmartData)*, Halifax, NS, Canada, 2018, pp. 9–14.

[81] Y. Zhang, J. Wen, and F. Mo, "The application of Internet of Things in social network," in *2014 IEEE 38th International Computer Software and Applications Conference Workshops*, Vasteras, Sweden, 2014, pp. 223–228.

[82] D. Hussein, S. N. Han, G. M. Lee, N. Crespi, and E. Bertin, "Towards a dynamic discovery of smart services in the social Internet of Things," *Computers & Electrical Engineering*, vol. 58, pp. 429–443, 2017.

[83] P. Garcia Lopez et al., "Edge-centric computing: Vision and challenges," *ACM SIGCOMM Computer Communication Review*, vol. 45, no. 5, pp. 37–42, 2015.

[84] L. Atzori, A. Floris, R. Girau, M. Nitti, and G. Pau, "Towards the implementation of the social internet of vehicles," *Computer Networks*, vol. 147, pp. 132–145, 2018.

[85] H. Z. Asl, A. Iera, L. Atzori, and G. Morabito, "How often social objects meet each other? Analysis of the properties of a social network of IoT devices based on real data," in *2013 IEEE Global Communications Conference (GLOBECOM)*, Atlanta, GA, USA, 2013, pp. 2804–2809.

[86] C. Turcu and C. Turcu, "The social internet of things and the RFID-based robots," in *2012 IV International Congress on Ultra Modern Telecommunications and Control Systems*, St. Petersburg, Russia, 2012, pp. 77–83.

[87] S. R. Islam, D. Kwak, M. H. Kabir, M. Hossain, and K.-S. Kwak, "The Internet of Things for health care: A comprehensive survey," *IEEE Access*, vol. 3, pp. 678–708, 2015.

[88] V. Miori and D. Russo, "Improving life quality for the elderly through the Social Internet of Things (SIoT)," in *2017 Global Internet of Things Summit (GIoTS)*, Geneva, Switzerland, 2017, pp. 1–6.

[89] P. Hao and X. Wang, "A PHY-aided secure IoT healthcare system with collaboration of social networks," in *2017 IEEE 86th Vehicular Technology Conference (VTC-Fall)*, Toronto, ON, Canada, 2017, pp. 1–6.

[90] N. Gulati and P. D. Kaur, "FriendCare-AAL: A robust social IoT based alert generation system for ambient assisted living," *Journal of Ambient Intelligence and Humanized Computing*, vol. 13, no. 4, pp. 1735–1762, 2022.

[91] J. A. Guerrero-Ibanez, S. Zeadally, and J. Contreras-Castillo, "Integration challenges of intelligent transportation systems with connected vehicle, cloud computing, and Internet of Things technologies," *IEEE Wireless Communications*, vol. 22, no. 6, pp. 122–128, 2015.

[92] M. Whaiduzzaman, M. Sookhak, A. Gani, and R. Buyya, "A survey on vehicular cloud computing," *Journal of Network and Computer Applications*, vol. 40, pp. 325–344, 2014.

[93] O. Kaiwartya et al., "Internet of vehicles: Motivation, layered architecture, network model, challenges, and future aspects," *IEEE Access*, vol. 4, pp. 5356–5373, 2016.

[94] B. L. R. Stojkoska and K. V. Trivodaliev, "A review of Internet of Things for smart home: Challenges and solutions," *Journal of Cleaner Production*, vol. 140, pp. 1454–1464, 2017.

[95] P. Rashidi, D. J. Cook, L. B. Holder, and M. Schmitter-Edgecombe, "Discovering activities to recognize and track in a smart environment," *IEEE Transactions on Knowledge and Data Engineering*, vol. 23, no. 4, pp. 527–539, 2010.

[96] A. Floris, S. Porcu, L. Atzori, and R. Girau, "A Social IoT-based platform for the deployment of a smart parking solution," *Computer Networks*, vol. 205, p. 108756, 2022.

[97] A. Piras, S. Mirri, M. Sole, D. Giusto, G. Pau, and R. Girau, "Implementation of a sea monitoring system based on social Internet of Things," in *2022 IEEE 19th Annual Consumer Communications & Networking Conference (CCNC)*, Las Vegas, NV, USA, 2022, pp. 687–690.

[98] Y.-K. Chen, "Challenges and opportunities of Internet of Things," in *17th Asia and South Pacific Design Automation Conference*, Sydney, NSW, Australia, 2012, pp. 383–388.

[99] S. Ali, M. G. Kibria, M. A. Jarwar, H. K. Lee, and I. Chong, "A model of socially connected web objects for IoT applications," *Wireless Communications and Mobile Computing*, vol. 2018, 20 pages, 2018.

[100] W. Mardini, Y. Khamayseh, M. B. Yassein, and M. H. Khatatbeh, "Mining Internet of Things for intelligent objects using genetic algorithm," *Computers & Electrical Engineering*, vol. 66, pp. 423–434, 2018.

[101] K. Venugopal, K. Srinivasa, and L. M. Patnaik, *Soft Computing for Data Mining Applications*, Springer, Berlin Heidelberg, Germany, 2009.

[102] M. Nitti, L. Atzori, and I. P. Cvijikj, "Network navigability in the social Internet of Things," in *2014 IEEE World Forum on Internet of Things (WF-IoT)*, Seoul, Korea (South), 2014, pp. 405–410.

CHAPTER 10

The Role of Software Defined Internet of Things (SDIoT) in Cloud Computing

Hania Batool and Adila Mehdi

BUITEMS

Ahthasham Sajid

Capital University of Science and Technology

10.1 BACKGROUND

Software-Defined Networks (SDN) and the Internet of Things (IoT) are two cutting-edge technologies. SDN provides network management orchestration by isolating the control plane from the data plane, while the Internet of Things tries to connect things through the Internet. There are billions of linked objects, making governing and controlling these difficult over an extensive dispersed network. SDN adds programmability and flexibility to the IoT network without interfering with existing solutions' underlying architecture. The IoT will not exist without software-defined networks. SDN virtualizes IoT networks at a cheap cost, enabling autonomous bandwidth allocation, device reconfiguration, and traffic rerouting to increase efficiency and simplify operations [1].

In recent years, cloud computing has been considered one of the most recent and emerging computing systems. Cloud computing, like numerous other technologies, originated from methodologies utilized by distributed systems and utility computing [2]. Distributed systems enabled capabilities like scalability, parallelism, continuous availability, heterogeneity, and fault isolation to fulfill the requirement of efficient and effective exploitation of shared resources across networks. Despite this, issues such as poor resource allocation and use across various administrative domains or organizations occurred [2]. To tackle this problem, the concept of grid computing was proposed, with the goal of offering a framework that solves the challenges that traditional distributed systems face in regard to scalability, resource heterogeneity, collaboration, policy-based management, and dynamic

DOI: 10.1201/9781032648309-13

provisioning [3]. Cloud computing is frequently referred to as the "successor of grid computing" [4]. Virtualization was launched about 40 years back as a result of advances in research in collaborative computing. It refers to the process of constructing a virtual layer on top of hardware that enables users to run numerous instances on the hardware at the same time. Virtualization is a fundamental cloud computing technology that serves as a foundation for major cloud services such as EC2 from Amazon [5]. Virtualization pioneered the way with functions that operated as intermediaries.

10.2 INTRODUCTION

Before we begin the survey of SDN in cloud computing, we explain the fundamental history of SDN in Section 10.2.1, the addition of SDN in cloud computing in Section 10.2.2, the basic architecture of an SDN network in Section 10.2.3, and example applications of SDN in cloud computing in Section 10.2.4. Section 10.3 summarized related works, while Section 10.4 gave a survey study of them. Section 10.5 discusses this chapter conclusion and potential future research.

10.2.1 History of Software-Defined Networking (SDN)

Following the boom in surfing the web in the 1990s, network operators and researchers worked together to make existing networks more manageable by making them programmable. As a consequence, the Defense Advanced Research Projects Agency (DARPA) of the United States developed the concept of Active Networking in the mid-1990s, which gave an innovative programming interface to manage a network. This was a "clean slate" approach since it addressed the challenges that the network's operators encountered at the time [6]. It allowed network designers to create control planes, which are the control software and network hardware components that are used to manage the behavior of the network transport system. Although active networking was a promising concept because it reduced computing costs, enabled network experimentation, and allowed service providers to deploy new network services as needed, it did not garner much interest, was not adopted, and wasn't worked on extensively because it did not provide a pragmatic and compelling deployment path [7].

During the unprecedented surge in Internet traffic volume in the early 2000s, network performance and predictability necessitated numerous changes. Network debugging setup issues, as well as regulating and forecasting network routing behavior, become extremely difficult. This was owing in part to conventional switches and routers' rigorously specified interfaces and protocols across the control and data planes. To address these issues, researchers undertook research on separating the data and control planes, which was even recommended in active networking models [8]. Recent development has mostly focused on rendering the control plane more programmable, rather than on the data plane, in order to make network managers' lives easier. ForCES (Forwarding and Control Element Separation), an open interface method put forth by the Internet Engineering Task Force (IETF) [9], standardized the simplicity of management between the data and control planes. The industry's demands for protocols and technologies to manage the expanding breadth and size of the networks they operate with ease led to more research on the separation of data and control plane, which in turn proved to have served as a stimulus for that research. Although it solved many of the issues that the network operators were facing at the time, ForCES along with other APIs were not widely implemented because they placed significant limitations on the amount of functionality that a programmable controller could provide. It did not provide a wide variety of capabilities including altering, flooding, and discarding packets for an extensive variety of protocols (like TCP and UDP protocols) and header field contents (like IP and MAC addresses). Further innovative techniques were however made possible by the concept of splitting the control and data planes through an open interface [7].

Notably, the Ethane project worked on a logically centralized controller that would maintain the global network policy that would regulate all incoming and outgoing packets in the mid-2000s, further investigating the control plane interface [10]. It is composed of switches that connect to the central controller and store basic flow tables. Upon receiving data packets, switches transmit them to the controller, which has the authority to authorize them. The establishment of the current OpenFlow API was made possible by the accomplishments of the Ethane project. OpenFlow enabled fully programmable networks and presented a workable solution for enabling real-world implementation, thereby addressing the shortcomings and constraints of the previous approaches. By adding new features and enhancing the capabilities of the switch gear already in place, it made their lives easier [7]. The idea of implementing OpenFlow in an intranet on a college campus to improve its programming was first proposed by a group of researchers from Stanford University [11]. The study's objective was to determine the OpenFlow protocol's capabilities by testing it on networks that spanned several schools as well as a single campus. Software Defined Networking, often known as SDN, was gradually implemented in networks other than campus networks, like a data center network, after the usefulness of OpenFlow was seen. When network designers realized that it was less expensive to source proprietary network software for commodity switches than to buy custom switches that required to be frequently renewed since they couldn't run the newest capabilities offered by suppliers, SDN reduced operating costs. SDN is the required paradigm that provides a programmable interface through which developers can construct software that can manipulate routing and access control features [12].

10.2.2 SDN in Cloud Networks

Since its launch, cloud computing has been the actualized and concrete manifestation of resource sharing on demand. This new paradigm combines the ideas of virtualization, connection, processing power, and storage to share computing resources across the internet [13]. It is an established, dependable, and well-developed method that gives users access to servers, storage, processing power, apps, and services in accordance with their needs. So avoiding the enormous expenses and overheads that would arise from buying the entire set of hardware and processing power.

Many organizations, including universities (Teesside University, University of Sydney, and Middlesbrough College), the healthcare industry (National Health Service), the banking industry, a number of businesses (Google, IBM, Alibaba), and social media platforms

(Facebook), have adopted and invested in cloud computing due to its many benefits [14]. These companies changed from using traditional networking models to cloud computing models by leasing services from cloud service providers.

Cloud service providers oversee various enormous data centers with a vast number of servers connected to a network that is controlled by a multitude of switches and other devices [15]. Although these data center networks are dispersed among several physical locations, they are all connected to create a single cloud network. Every server makes available processing power in response to user requests.

Since so many switches and routers are utilized in data centers, setting up and maintaining these networks may be very complicated. Furthermore, because the control logic and data planes are grouped inside the devices, they function as separate autonomous systems and decide on their own routes [16]. Consequently, every switch and router on the network needs to have its configuration changed if any changes are to be made. This conventional networking method exacerbates the already expensive network design and damages the network as a whole.

To counteract the shortcomings of the conventional network method, cloud data center networks began implementing Software Defined Networking (SDN). SDN makes network administration easier by converting switches and routers into forwarding devices and eliminating the clustering data planes and control logic. As previously said, SDN implements a logically centralized controller that oversees all network forwarding elements, separating the control plane from the data plane. Since the controller is software-based, its underlying network functions are abstracted [17]. Because SDN is flexible and allows network operators to dynamically alter network traffic, it is ideal for cloud computing. SDN can be advantageous to a cloud computing architecture because it offers a global view of the cloud's data center network, dynamic workload balancing, and network security and virtualization through its programmable interface.

10.2.3 The Architecture of SDN Clouds

Let's start our explanation of SDN in cloud networks with a quick rundown of its design, as shown in Figure 10.1 below. A variety of SDN-enabled cloud architectures, like Meridian, have been proposed in numerous research articles [17–19]. Nevertheless, we shall outline a condensed form of the entire architecture, which is made up of the following parts:

10.2.3.1 Network Application Layer

The network application layer, represented as the top layer in Figure 10.1, is home to the SDN applications and modules that either directly interact with the network or control it via the controller. It does access control, traffic management, optimum path configuration, debugging operations, and features and applications like Volume Based Billing (VBB), Volume Based Controllers (VBC), Service Load Balancers (SLB), Firewall (FW), and Quality of Service (QoS) nodes. Additionally, it serves as a conduit via which any business can use an application provisioning process to automatically assign suitable network settings, like firewall traffic rules [18].

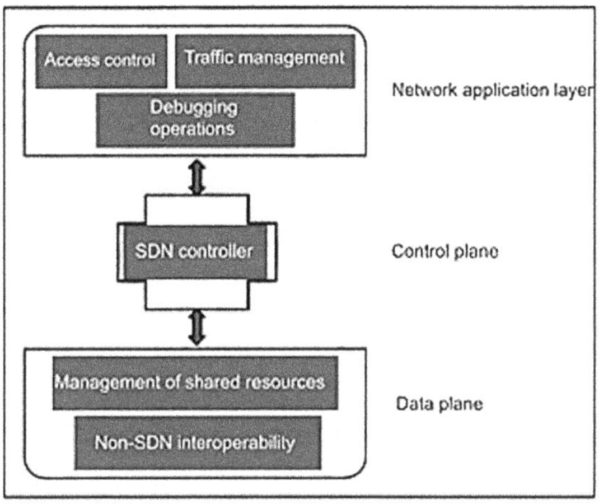

FIGURE 10.1   Architecture of an SDN network.

10.2.3.2 Control Plane

The SDN controller makes up the control plane. As illustrated in Figure 10.1, it is the logically centralized controller previously discussed that sits beneath the network application layer. Applications are served by the controller, which computes paths for packets to follow based on their routing algorithms, network-wide topology views, and device configuration. It also maintains each node informed of any changes in the availability or lack thereof of any network link and orchestrates the various commands generated by the applications [18]. It also does a number of other tasks after receiving packets from the layer above, one of which is converting requests and instructions from the logical format created by the application layer into a physical format. Since the apps use APIs to communicate with the network, it is essential to appropriately translate these high-level commands for the underlying network. Usually, the OpenFlow [11] protocol is used to enable the controller to communicate with network devices. Some of the SDN controllers that are commonly used and accessible are Beacon [20], OpenDayLight [21], Floodlight [22], and NOX.

10.2.3.3 Data Plane

Through the last layer as presented in Figure 10.1 previously, which functions as an interface made possible by drivers, plugins, switches, and routers, the controller can communicate with network tools [23]. This layer manages shared resource management, interoperability functions and decisions with non-SDN enabled networks, internal traffic processing, external traffic forwarding, and shared resource management. The controlled utilized such plugins and drivers performs the tasks of building, evaluating, and upgrading the network through a global view, as well as assigning instructions to specific devices and gathering network information [18]. Additionally, in a network with OpenFlow enabled, a controller can configure the insertion rules for a switch. Every OpenFlow-based switch in the network keeps flow tables up to date. It contrasts data packet header fields with those found in its flow tables. If the header fields of the packets match, the switch chooses to route them to the appropriate locations. If the header fields of the packets are not identical, the switch returns the packets to the controller, which decides whether to discard them or create a new match field that is specific to each packet [24].

10.2.4 Application of SDN in Cloud Computing

Because SDN makes cloud networking possible, it has been employed in a variety of applications. The following section has covered a few of such facets:

10.2.4.1 Security

Distributed Denial of Service (DDoS) attacks are the most dangerous ones when it comes to seriously jeopardizing the security of SDN-enabled clouds. The network is unable to provide regular client service due to the enormous volume of heavy traffic. A Distributed SDN Controller was suggested as a solution to this issue. A vulnerability study was conducted to protect real-time cloud data centers against DDoS attacks. For the purpose of this investigation, Open Daylight (ODL) and Open Networking Operating System (ONOS), two well-known SDN controllers, were used in this report. DDoS assaults were launched against the ONOS regulator and the ODL-3 node cluster regulator using various scripts. The ODL 3-node cluster was shown to be superior to the ONOS regulator [25]. Similarly, a hardware platform that speeds up packet processing in virtual switches and a security cluster-based SDN controller that monitors and controls cloud networks were built and put into place while testing the resilience and security of an SDN-enabled cloud. It was found that the SDN controller cluster enhanced network strength in the event of an assault, even in the face of extremely unusual network traffic. Switches with hardware acceleration perform well and fit in well with cloud systems. The highly accessible and secure SDN-enabled cloud computing approach utilized in the integration of Open Stack and Open Daylight was first presented in this research study [26].

Moreover, work on a DistB-SDCloud architecture was done to improve the security of IOT apps that are cloud-enabled. They suggested a BlockChain (BC) architecture to preserve and enable network security, integrity, privacy, and confidentiality. By combining SDN with BC, they came to the conclusion that response times, throughput, and CPU utilization measures are secure even in the face of network attacks [23].

10.2.4.2 Virtualization

Blue Bird proposed a sophisticated network virtualization technique with excellent performance for a minimally functional cloud service on Azure. In order to prevent severe performance loss as a result of increased demand or scaling up, Bluebird uses a few well-established specialized principles in the control plane, guaranteeing that the network stays resilient and fault-tolerant. By incorporating route caching techniques, generalizing and abstracting network interface devices, and severing the centralized controller's functions from the agents embedded in switches, these objectives are satisfied. The network design can now be more flexible, scalable, and fault-tolerant thanks to this decoupling. By observing data plane traffic, the research paper's authors demonstrated how route caching can reduce network latency [27].

10.2.4.3 SDN Controller Platform

In order to access cloud apps, users are typically required to establish a number of network layer structures, including switches, subnets, and Access Control Lists (ACLs). A cloud application network's service-level model is supported by Meridian, an SDN controller platform. By providing a way to effectively manage dynamic changes to the virtual network, coordinate network duties across a large number of devices, and interact with numerous cloud controllers, the architecture and implementation overcome those drawbacks [18].

10.2.5 The Importance of SDN in the Internet of Things

Software-defined networking, or SDN, is a major improvement over traditional networking in the context of the Internet of Things (IoT) and provides several important benefits [28].

10.2.5.1 Improved Control with Unmatched Speed and Flexibility

The need to manually configure a range of hardware components from various manufacturers is eliminated by SDN. Alternatively, network traffic can be managed by developers through the configuration of an open-standard software-based controller. This approach allows networking managers to select networking equipment with remarkable speed and flexibility by utilizing a centralized controller to interact with several hardware devices over a single protocol [28].

10.2.5.2 Network Architecture That Is Customizable

Network managers may centrally design network services and rapidly and easily assign virtual resources to change the network architecture by using SDN. This feature allows network managers to optimize data flow across the network based on requirements and to give priority to applications that demand more availability [28].

10.2.5.3 Robust Security

SDN in the Internet of Things offers total network visibility, which gives a comprehensive view of potential security flaws. As more and more intelligent devices connect to the Internet, SDN performs better than traditional networking in terms of security advantages. Operators can implement separate zones for devices with different configurations or swiftly isolate compromised devices to prevent viruses from propagating throughout the network [28].

10.3 LITERATURE REVIEW

The growing use of SDN-enabled cloud and cloud computing has been thoroughly researched and developed in the last few years. The purpose of this review of the literature is to give a broad overview of the academic publications and research projects that have been done on the development and history of SDN in cloud networks. A critical analysis has been conducted, as Table 10.1 illustrates.

TABLE 10.1   Critical Analysis of Literature Review

Title	Author	Contribution	Security	Virtualization	Scalability of SDN
A taxonomy of software-defined networking (SDN)-enabled cloud computing [15]	Jungmin Son and Rajkumar Buyya	A survey of SDN enabled cloud computing emphasizing on both the networking and distributed systems aspects of SDN and cloud computing	Yes	Yes	No
A survey on SDN, the future of networking [17]	Shiva Rowshanrad et al.	A survey on the use of SDN for cloud computing focusing on various SDN architectures, wireless architectures and different approaches used in SDN-enabled clouds	Yes	Yes	No
Cloud computing networking: Challenges and opportunities for innovations [29]	Siamak Azodolmolky et al.	A survey on cloud computing services centering on Infrastructure as a Service (IaaS) while describing the challenges and opportunities revolving around IaaS technologies. SDN-based cloud federation has also been explored	Yes	Yes	No
The road to SDN: An intellectual history of programmable networks [7]	Nick Feamster et al.	The history and direct evolution of programmable networks such as active networks, the control and data plane, and OpenFlow technology that laid the foundation for SDN have been discussed	No	Yes	No
SDN orchestration architectures and their integration with cloud computing applications [30]	Arturo Mayoral et al.	This study focused on the evolution of SDN. The target of this study is the review of several SDN controllers and their respective architectures	No	No	No

A complete and in-depth analysis of SDN usage studies for cloud computing was carried out by Jungmin Son and Rajkumar Buyya. The main foci of their analysis were network virtualization and security, data center energy efficiency, and network optimization for packet handling procedures. Furthermore, the QoS management techniques of an SDN network were outlined and expanded upon based on several research instruments for optimizing energy consumption, modeling, and utilization maximization. Their analysis, however, lacked the literature necessary to close the knowledge gap between the state of the art and cloud-optimized data centers, as well as their eventual autonomy [15].

Similarly, the control, infrastructure, and application layers are the three separate levels that Shiva Rowshanra et al. used to characterize the SDN architecture. Lastly, SDN research trends and applications were unveiled, including virtual data centers and mobile and wireless networks [17].

Additionally, Siamak Azodolmlky et al. discussed interconnection issues and network issues inIaaS, which are presently being resolved by current technology. This article focused on virtual networking and cloud interconnection to illustrate the architecture of Infrastructure as a Service (IaaS) and its increasing limitations. A collection of an SDN architecture's Application Programming Interfaces (APIs) was highlighted in order to illustrate how ordinary network functions might be made simpler. An SDN-based federation that makes it easier for customers and service providers to choose a good cloud data center for their needs was covered in the article. However, the scalability of this kind of network was not investigated [29].

Similar to this, Nick Feamster et al. covered the history and development of programmable networks, including active networks, control and data planes, and recent technological advancements that opened the door for SDN development, but they also covered all the pertinent information regarding the ideas that were put forth at the time and the workable solutions that made them a reality. The authors outlined a precise path that led from earlier iterations of network architectures to the creation of SDN. However, since some of the features and functions in OpenFlow are outdated and need to be improved, the limitations of the protocols were also covered, including the need to transparently examine problems before using them in SDN applications [7].

The evolution of SDN and contemporary logically centralized SDN controller technologies, like Generalized Multi-Protocol Label Switching (GMPLS), were the subject of Arturo Mayoral et al.'s study. GPLS necessitates coordination between multiple supervised and regulated networks in order to offer third-party vendors highly manageable and open connectivity solutions. To determine the possible benefits and drawbacks of the two designs, they specifically contrasted the Application-Based Network Operations (ABNO) architecture with a single SDN controller orchestration method. The problem of network coordination in multi-domain networks was thoroughly examined in this study [30], which covered a wide range of technologies in the transport and control planes.

10.4 SURVEY ANALYSIS

Network architectures have become simpler as a result of the advent of SDN, according to the analysis conducted for this study. The following are some of the talks of the examined aspect: managing big datasets and capable of building extended models to successfully solve the overfitting issue:

10.4.1 Increased Convenience

In contrast, there were few options for extending and growing a network in the past, and cloud network and IoT management tools were scarce. Creating and configuring a cloud network is now easier with the availability of programmable networks via interactive interfaces. In the past, in order to manage network load, specialized, custom switches had to be made, which resulted in a high overhead of having to adjust every network device whenever something changed. But nowadays, specialized software is created that allows for easy network reconfiguration and adapts those changes to the entire network.

10.4.2 Robust Security

SDN has improved a cloud network's security, virtualization capabilities, and performance while also making it much easier to plan, organize, coordinate, and govern network resources. High levels of malicious traffic and DDoS attacks can be controlled and stopped without endangering the cloud network's ability to operate continuously.

10.4.3 Minimized Complexity

By separating the control and data planes, the aforementioned achievements have become achievable and provide a transparent opportunity to simplify network architecture. SDN de-clutters the complex web of network devices and protocols, multiple clients and their differing needs, troubleshooting and configuration methods, network testing, and final installations.

10.4.4 Research Gaps

But, additional investigation is required into the possibility of scalability for an SDN-enabled cloud-based Internet of Things network. Because of the growing number of users connecting to cloud networks and the increased need for resource sharing, cloud networks must be readily scalable. Since any technical approach has drawbacks, further research and testing are necessary to determine how scalable an SDN cloud architecture can be.

10.5 CONCLUSION AND FUTURE WORK

SDN is a fairly innovative network architecture on which numerous research and studies have been conducted due to its novel approach to converting traditional networks into interactive programmable ones. However, there are still some aspects of SDN in cloud and IOT networks that require more in-depth exploration such as experimenting with the resilience of its security, its performance when operating with a non-SDN enabled cloud network, its scalability with respect to the exponentially growing need for on-demand resource sharing, and different implementations and architectures of SDN enabled clouds should be compared and analyzed. Additionally, the use of Artificial Intelligence (AI) and machine learning in an SDN-enabled cloud-based IOT networks could be further explored. With the recent advancements in AI, cloud networks should also benefit from the opportunities made possible by AI algorithms. Hands-free features such as automatically selecting relevant network protocols for an enterprise, self-regulating traffic management, autonomous energy consumption detection mechanisms, and mechanized diagnostic attributes are only some of the potential future research directions.

In this article, SDN used in cloud computing and IOT networks have been reviewed and discussed. The essence of this article is the history of SDN and its use in cloud data centers that form the cloud networks along with the architecture of SDN through its three layers namely network application layer, data plane layer, and control plane layer. Additionally, the applications of SDN in cloud computing and features for IOT networks are also mentioned, such as SDN security, virtualization, and various control platforms.

REFERENCES

[1] Y. İnağ, M. Demirci, and S. Özemir, "Implementation of an SDN based IoT network model for efficient transmission of sensor data," *2019 4th International Conference on Computer Science and Engineering (UBMK)*, Samsun, Turkey, 2019, pp. 682–687, doi:10.1109/ UBMK.2019.8907119.

[2] R. P. Padhy and M. R. Patra, "Evolution of cloud computing and enabling technologies," *International Journal of Cloud Computing and Services Science (IJ-CLOSER)*, vol. 1, no. 4, pp. 183–191, 2012.

[3] J. Joseph, M. Ernest, and C. Fellenstein, "Evolution of grid computing architecture and grid adoption models," *IBM Systems Journal*, vol. 43, no. 4, pp. 624–645, 2004.

[4] R. Kaur and A. Kaur, "A review paper on evolution of cloud computing, its approaches and comparison with grid computing," p. 6062, Accessed: July 06, 2023. [Online]. https://citeseerx.ist.psu.edu/document?repid=rep1&type=pdf&doi=7d94150feaeeddd1cba7f1bd8ca92a50f c0efb3c

[5] Y. Xing and Y. Zhan, "Virtualization and cloud computing," In: Zhang, Y. (eds) *Future Wireless Networks and Information Systems. Lecture Notes in Electrical Engineering*, vol 143. Springer, Berlin, Heidelberg, p. 306, 2012.

[6] K. Calvert, "Reflections on network architecture," *Computer Communication Review*, vol. 36, no. 2, p. 27, 2006.

[7] N. Feamster, J. Rexford, and E. Zegura, "The road to SDN," *ACM SIGCOMM Computer Communication Review*, vol. 44, no. 2, pp. 2–4, 2014.

[8] J. M. Smith, K. L. Calvert, S. L. Murphy, H. K. Orman, and L. L. Peterson, "Activating networks: A progress report," *Computer*, vol. 32, no. 4, pp. 32–41, 1999.

[9] L. Yang, R. Dantu, T. J. Anderson, and R. Gopal, "Forwarding and control element separation (ForCES) framework," 2004. https://www.rfc-editor.org/rfc/rfc3746.html

[10] M. Casado, M. Freedman, J. Pettit, J. Luo, N. Mckeown, and S. Shenker, "Ethane: Taking control of the enterprise," *ACM SIGCOMM computer communication review*, *37*(4), pp. 1–12, 2023.

[11] N. McKeown et al., "OpenFlow," *ACM SIGCOMM Computer Communication Review*, vol. 38, no. 2, p. 70, 2008.

[12] T. Koponen et al., "Onix: A distributed control platform for large-scale production networks," In *9th USENIX Symposium on Operating Systems Design and Implementation (OSDI 10)*, Vancouver, BC, Canada, pp. 351–364, 2010.

[13] M. Malathi, "Cloud computing concepts," *2011 3rd International Conference on Electronics Computer Technology*, Kanyakumari, India, 2011, pp. 236–239.

[14] L. Golightly, V. Chang, Q. A. Xu, X. Gao, and B. S. Liu, "Adoption of cloud computing as innovation in the organization," *International Journal of Engineering Business Management*, vol. 14, p. 18479790221093992, 2022.

[15] J. Son and R. Buyya, "A taxonomy of software-defined networking (SDN)-enabled cloud computing," *ACM Computing Surveys*, vol. 51, no. 3, pp. 59.2–69.4, 2018.

[16] T. Zhang, "Distributed controllers in software defined networks", pp. 5–6, 2014.

[17] S. Rowshanrad, S. Namvarasl, V. Abdi, M. Hajizadeh, and M. Keshtgary, "A survey on SDN, the future of networking," *Journal of Advanced Computer Science & Technology*, vol. 3, no. 2, p. 232, 2014.

[18] M. Banikazemi, D. Olshefski, A. Shaikh, J. Tracey, and G. Wang, "Meridian: an SDN platform for cloud network services," *IEEE Communications Magazine*, vol. 51, no. 2, pp. 120–127, 2013.

[19] Y.-H. Chu, Y.-T. Chen, Y.-C. Chou, and M.-C. Tseng, "A simplified cloud computing network architecture using future internet technologies," *International Journal of Networks and Communications*, vol. 2, no. 5, pp. 105–111, 2012.

[20] D. Erickson, "The beacon openflow controller," *Proceedings of the Second ACM SIGCOMM Workshop on Hot Topics in Software Defined Networking (HotSDN '13)*. Association for Computing Machinery, New York, pp. 13–18, 2013.

[21] OpenDayLight, 2013. [Online]. Available: https://www.opendaylight.org/.

[22] "Floodlight," 2012. Available: https://floodlight.openflowhub.org.

[23] A. Rahman, M. J. Islam, S. S. Band, G. Muhammad, K. Hasan, and P. Tiwari, "Towards a blockchain-SDN-based secure architecture for cloud computing in smart industrial IoT," *Digital Communications and Networks*, vol. 9, no. 2, pp. 411–421, 2022.

[24] K. Govindarajan, K. C. Meng and H. Ong, "A literature review on Software-Defined Networking (SDN) research topics, challenges and solutions," *2013 Fifth International Conference on Advanced Computing (ICoAC)*, Chennai, 2013, pp. 293–299.

[25] S. Badotra et al., "A DDoS vulnerability analysis system against distributed SDN controllers in a cloud computing environment," *Electronics*, vol. 11, no. 19, p. 3120, 2022.

[26] L. T. Le and T. N. Thinh, "Enhancing security and robustness for SDN-EnabledCloud networks," *REV Journal on Electronics and Communications*, vol. 12, no. 1–2, p. 8, 2021.

[27] M. Arumugam, D. Bansal, N. Bhatia, and J. Boerner, "Bluebird: High-performance SDN for bare-metal cloud services," *USENIX*, pp. 2–10, 2022.

[28] S.K. Tayyaba et al., "Software defined network (SDN) based internet of things (IOT)," *Proceedings of the International Conference on Future Networks and Distributed Systems* [Preprint], Cambridge, United Kingdom, 2017. doi:10.1145/3102304.3102319.

[29] S. Azodolmolky, P. Wieder, and R. Yahyapour, "Cloud computing networking: Challenges and opportunities for innovations," *IEEE Communications Magazine*, vol. 51, no. 7, pp. 54–62, 2013.

[30] A. Mayoral, R. Vilalta, R. Muñoz, R. Casellas, and R. Martínez, "SDN orchestration architectures and their integration with cloud computing applications," *Optical Switching and Networking*, vol. 26, pp. 2–13, 2017.

CHAPTER 11

Internet of Vehicles (IoV)

Challenges, Threats and Routing Protocols

Mariya Ouaissa

Cadi Ayyad University

Mariyam Ouaissa

Chouaib Doukkali University

Soukayna Riffi Boualam

Moulay Ismail University

Zakaria Boulouard

Hassan II University

Inam Ullah Khan

Szabist University

Sarah El Himer

Sidi Mohamed Ben Abdellah University

11.1 INTRODUCTION

Recently, Internet of Vehicles (IoV) occupies an important place in research fields. This research is linked with the development of the automotive industry and wireless communication technologies. Nowadays, the use of vehicles is increasing rapidly, and this can lead to road blockages. Consequently, the problem of traffic congestion intervenes and creates an imbalance in the flow of traffic, especially when vehicles circulate in specific areas during peak hours. At that time, the number of cars exceeds the maximum capacity of the roads. To avoid traffic congestion, researchers have developed connected vehicles to

DOI: 10.1201/9781032648309-14

improve the action plans present in the automotive industry, and they also improve travel by choosing the most optimal routes. Several researches on data processing and security have been made in order to choose the best decisions to improve traffic in urban areas. The use of the IoV has made it possible to control contact between vehicles and with the outside world; thus, it plays an important role when collecting and processing data in an environment. Quality of Service (QoS) is mandatory in the design of connected vehicles [1,2].

This chapter is organized into the following sections. In Section 11.2, we present an overview of IoV. Section 11.3 presents the challenges faced by connected vehicles. Section 11.4 discusses IoV routing protocols. Section 11.5 illustrates the design of some architectures that guarantee security and reliability during a communication exchange. In Section 11.6, we describe the performance of using RPL under IoV system. We conclude in Section 11.7.

11.2 INTERNET IN CONNECTED VEHICLES

The Internet of Vehicles is developed based on the Internet of Things (IoT). The IoT is considered as a network of objects that are connected to the internet in order to obtain communication and transfer information. The connected objects platform has become famous and can be used in several areas of development and research. Among these domains, we find the domain that focuses on IoV network. The latter processes vehicles that communicate with each other in different infrastructures. Subsequently, this type of vehicle is developed toward connected autonomous smart vehicles. IoV makes it possible to facilitate the infrastructure of intelligent vehicles and thus makes it possible to drive easily without any human intervention by taking into account the security and the reliability of the communication. It even allows you to analyze the road. Connected vehicles focus on research and data collection using specific sensors that are embedded in connected vehicles. The collected data deal with several important points like speed, location, road flow, and direction of movement.

Vehicle networks operate by following a layered architecture. The goal is to meet several needs such as reliability, security, flexibility, and interoperability. The layered architecture proposed by several researchers can be divided into three essential layers: A layer that combines the physical layer and the data link layer, which is called the detection layer; the network layer; and the application layer [3,4]

The first layer is the one at the bottom of the architecture. The latter is responsible for collecting data using specific sensors. These data include the way drivers drive their vehicles, and the state and flow of the environment. This layer initially makes it possible to detect and collect data from vehicles present in the environment. Subsequently, these data are converted into digital signals. And finally, the data are sent to the network layer. Among the technologies that make it possible to detect information, we cite Radio Frequency Identification (RFID) tags and wireless sensor networks. The second layer is the network layer used to process data and transmit it to the upper layer. It also provides connectivity and manages communication between vehicles. The third and last layer is the application layer which provides storage and data processing and decision-making. This layer is

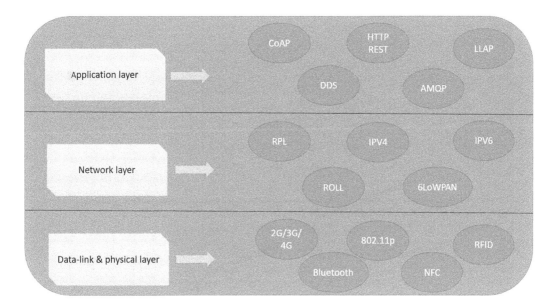

FIGURE 11.1   The layered architecture of connected vehicles.

responsible for the management of the applications, the management of the final data, and their security during the communication (Figure 11.1).

In recent years, the IoV platform improves transport and safety, reduces fuel use, and also reduces the number of accidents on the road. Several researches are in progress, so that this system is universal and it is integrated between hardware technologies and software applications.

11.3 INTERNET OF VEHICLES CHALLENGES

Connected vehicles suffer from several problems, several challenges are put in front of this type of vehicle. During the design of this type of vehicle, it is necessary to take into account the variation in the number of connected devices in the vehicle, the state of the environment or the road with regard to the flow of traffic which is generally unbalanced, the reliability of the connection which allows the data to be disseminated in an optimal time [5]. In this fact, the major challenge of IoV is to disseminate data from one vehicle to another while ensuring reliability. Transferring information with reliability in a short time allows real-time applications to be implemented for the IoV environment. These applications allow you to select the optimal communication paths in specific environments.

Also, while moving, vehicles in IoV are restricted to road topology, if traffic information is available, it is possible to predict the current and future position of a vehicle, detect nearby vehicles, and provide communication and continuous information transfer without interruption.

IoV suffers from several difficulties, such as the mobility of the latter is high. Then, the nodes are much more dynamic; the vehicles move with great speed and change their positions at every moment. This frequency of movement makes it possible to obtain a dynamic

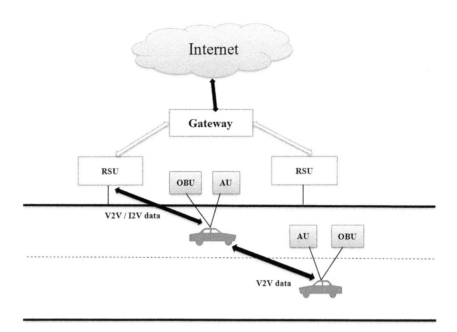

FIGURE 11.2   Internet of vehicles architecture.

network topology. However, this high mobility often leads to loss of connection between nodes [6] (Figure 11.2).

11.4 ROUTING PROTOCOLS IN IoV

Routing is a mechanism that allows data packets to be transferred from one network to another. Routing in IoV differs from mobile network routing; the routing algorithms that occur in mobile are not available for application scenarios in vehicle networks. The big challenge is to find the best architecture to reduce the delay when transferring data from one node to another while ensuring good quality of service. Furthermore, the routing protocol must take into account the nature of the vehicle network topology, and it must be able to overcome the obstacles of certain architectures and operate in a dynamic topology that supports irregular connectivity. IoV is based on wireless technology to ensure communication between different services and applications [7,8]. Topology-based protocols are classified into three categories: proactive, reactive, and hybrid.

For the proactive protocol, the routing table is updated by the nodes by inserting the information of new routes into the network. Hello packets are sent periodically to transfer data to neighboring nodes. This effect creates substantial control overhead and limits the use of available bandwidth.

For the reactive approach, updates are sent only when needed, which reduces control overhead substantially. However, this approach still has overhead such as route maintenance. The overhead created in the reactive protocols helps to discover the paths to send the information.

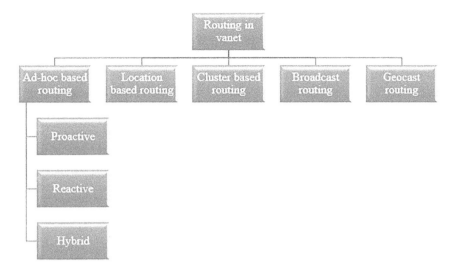

FIGURE 11.3   Routing protocols types in IoV.

Hybrid protocols are considered as a new innovation and discovery made by researchers. This type of protocol is analyzed in detail in Ref. [9]. This approach focuses on network design architecture more than performance analysis and improvement (Figure 11.3).

11.5 SECURITY AND PRIVACY FOR IoV

The security and confidentiality of connected vehicles are very important to ensure communication and data transfer. Researchers are still conducting research in order to overcome these security issues. Indeed, the data and messages circulating in the IoV must not be known by malicious servers and hackers in order to guarantee the integrity and confidentiality of the information transferred. All smart in-vehicle applications in connected vehicles must be protected against all malware [10–12].

In order to find a solution to this type of security and privacy issues precisely, in recent years, several approaches have been listed in the literature. These approaches deal with two aspects, namely, the exchange of communication and confidentiality [13]; the researchers worked on the security architecture that plays a key role in order to guarantee the security of vehicles. A low level of security can lead to false requests on road cloud services. In Ref. [14], the author describes the security architecture in view of the functional layer, the organizational view, the information extraction view, and the reference model view. Reference [15] focuses on improving the security of the wireless part in IoV and improving the confidentiality of communicators. The improved architecture is based on secure communication schemes and algorithms [12]. In Refs. [16,17], the authors present a communication architecture based on shared session keys; this type of key ensures secure communication over a long period. In Ref. [18], it is an improvement of Refs. [16,17], as the latter uses session keys in insecure applications. The latter uses two session keys: pair keys and group keys. In Ref. [19], this study is based on the use of VPKI (vehicular public key infrastructure), group signing, and CA (certificate authority). In Ref. [20], the actors proposed the use of several certified public keys instead of a single key. In Ref. [21], the author illustrates a security method that requires nodes to provide secure sender authentication even if the number of vehicles is large and the human factor exists in the environment. This study emphasizes obtaining node authentication trust.

11.6 PERFORMANCE OF USING RPL UNDER MOBILITY FOR IoV

The IoT presents a new technology that integrates all the connected devices in the world. Mobile vehicles as in the Internet network are considered as a new area called Internet of Vehicles. This technology combines the Internet of Things and vehicular networks. This technology presents a great challenge to ensure data communication between nodes. Several works are focused on designing an efficient solution that deals with routing problems in IoV.

Due to the mobility of IoV, it is mandatory to use routing protocols to transfer information and ensure a reliable exchange during communication between vehicles. Based on data exchange techniques, there are routing protocols that are either topology or position based. Topology-based routing protocols use path information to transmit data packets. This type of protocol is divided into reactive protocols, which are used on demand, and proactive protocols, which are routing table driven.

For IoV that are highly dynamic, position-based routing techniques are the most suitable to use. But, the problem that arises is that these techniques generate high costs because of the updates that are made continuously to transmit new information on the position and speed of vehicles [4]. Indeed, the protocols that will be used for IoV must take into consideration several vectors such as mobility, speed, data traffic, vehicular network density, and road topology.

Some work is done to test the performance of the RPL protocol on dynamic IoV environments. This distance vector routing protocol is based on IPv6, it focuses on low-power lossy networks (LLN) [22] such as wireless sensor networks. RPL constructs Destination Oriented Acyclic Graphs (DODAG) based on the objective functions and link and link metrics. Each DODAG follows a route in order to arrive at the root by traversing a tree topology [23]. The route is chosen using the objective function to optimize the cost. There are OFs that are based on a single metric, just as there are other OFs that combine several routing metrics. These metrics can be number of hops, number of expected transmissions (ETX), energy, latency, QoS, etc.

The most suitable routing protocol for this type of technology is the Routing Protocol for Low Power and Lossy Networks "RPL." RPL is an IPV6 routing protocol designed primarily for Wireless Sensor Networks (WSN). This is because the IPv6 protocol supports a large address space, hence it is considered the best for IoT. The IETF has standardized the RPL protocol, it is suitable for low power and lossy networks. The RPL protocol is based on an Objective Function that adapts to the requirements of the applications as needed. OF is a function that takes link and link metrics into consideration to choose the optimal route to the destination. Vehicular networks are mobile; this mobility allows flexibility and the integration of several services. On the other hand, this mobility causes problems of continuous disconnection which leads to the loss of data and transmission delays which are long.

RPL is suitable for dynamic network topology. And since IoV have high mobility, they suffer from frequent link disconnections. Also, the network density is high. These characteristics make data communication very difficult in the IoV. IoV communication is vehicle-to-vehicle (V2V) or vehicle-to-infrastructure (V2I). For this, we must find an effective way to make communication reliable by ensuring security, confidentiality, and data storage.

The reason the RPL protocol is the best to use for IoV is that its design is easy to modify to match the rate at which the parent node is updated. The RPL protocol is characterized by a better response time compared to other existing protocols. It is also characterized by high road availability. Another advantage of the RPL protocol is that it only requires the local information of the neighboring nodes to exchange the information, it does not take into account the global information on the network topology. So, the RPL protocol is the most suitable protocol for large-scale networks such as IoV.

11.7 CONCLUSION

This chapter presents a set of problems suffered by the Internet of Vehicles; thus, some research challenges are presented in relation to the security and confidentiality of vehicular communication. Also, it presents the architecture and routing protocols used to ensure communication between IoV. Several approaches carried out at the IoV level are cited in order to improve the security of communication exchanges. In addition, when it comes to data reporting, IoV presents several challenges. This is because this technology is characterized by several constraints such as high mobility, high network density, dynamic topology, and frequent disconnections of links. The advantage of the RPL protocol is that it is suitably modifiable so that it can work in a vehicular network. After some studies, it is clear that the RPL protocol provides better packet forwarding in vehicular networks.

REFERENCES

[1] S. R. Boualam, M. Ouaissa, M. Ouaissa, and A. Ezzouhairi, "Internet of connected vehicles: Challenges and opportunities," *AIP Conference Proceedings*, vol. 2814, no.1, pp. 1–6, 2023.

[2] M. Ouaissa, M. Ouaissa, M. Houmer, S. El Hamdani, and Z. Boulouard, "A secure vehicle to everything (V2X) communication model for intelligent transportation system," In Mariya Ouaissa, Zakaria Boulouard, Mariyam Ouaissa, and Bassma Guermah (eds.). *Computational Intelligence in Recent Communication Networks*, Cham: Springer International Publishing, 2022, pp. 83–102.

[3] N. Liu, "Internet of vehicles: Your next connection," *Huawei WinWin*, vol. 11, pp. 23–28, 2011.

[4] K. Golestan et al., "Situation awareness within the context of connected cars: A comprehensive review and recent trends," *Information Fusion*, vol. 29, pp. 68–83, 2016.

[5] M. D. Felice, I. V. Calcagni, F. Pesci, F. Cuomo and A. Baiocchi, "Self-healing infotainment and safety application for VANET dissemination," In *IEEE ICC*, London, UK, 2015.

[6] S. Zeadally, R. Hunt, Y.-S. Chen, A. Irwin, and A. Hassan, "Vehicular ad hoc networks (VANETS): Status, results, and challenges," *Telecommunication Systems*, vol. 50, no. 4, pp. 217–241, 2012.

[7] W. Liang, Z. Li, H. Zhang, S. Wang, and R. Bie, "Vehicular ad hoc networks: Architectures, research issues, methodologies, challenges, and trends," *International Journal of Distributed Sensor Networks*, vol. 5, p. 745303, 2015.

[8] M. S. Talib, B. Hussin, and A. Hassan, "Converging VANET with vehicular cloud networks to reduce the traffic congestions: A review," *International Journal of Applied Engineering Research*, vol. 12, no. 21, pp. 10646–10654, 2017.

[9] V. Nundloll, G. S. Blair, and P. Grace, *"A component-based approach for (re)-configurable routing in VANETs,"* In *Proceedings of the 8th International Workshop on Adaptive and Reflective Middleware*, Urbana Champaign, Illinois, December 2009.

[10] S. ur Rehman, "Vehicular ad-hoc networks (VANETs) – An overview and challenges," *Journal of Wireless Networking and Communications*, vol. 3, no. 3, pp. 29–38, 2013.

[11] F. Dotzer, *"Privacy issues in vehicular ad hoc networks,"* In *Proceedings of the 5th International Workshop on Privacy Enhancing Technologies (PET '05)*, Cavtat, Croatia, 2005, pp. 197–209.

[12] J. M. de Fuentes, A. I. Gonzlez-Tablas, and A. Ribagorda, *Overview of Security Issues in Vehicular Ad-Hoc Networks*, In: *Handbook of research on mobility and computing: Evolving technologies and ubiquitous impacts*. IGI global, pp. 894–911, 2011.

[13] F. Kargl, L. Buttyan, D. Eckhoff, P. Papadimitratos, and E. Schoch, *Working Group on Security and Privacy*, Karlsruhe Institute of Technology, Karlsruhe, Germany, 2011.

[14] M. Gerlach, A. Festag, T. Leinmller, G. Goldacker, and C. Harsch, *"Security architecture for vehicular communication,"* In *Proceedings of the 5th International Workshop on Intelligent Transportation (WIT '07)*, Hamburg, Germany, 2007.

[15] P. Papadimitratos, L. Buttyan, T. Holczer et al., "Secure vehicular communication systems: Design and architecture," *IEEE Communications Magazine*, vol. 46, no. 11, pp. 100–109, 2008.

[16] M. Raya and J.-P. Hubaux, "Securing vehicular ad hoc networks," *Journal of Computer Security*, vol. 15, no. 1, pp. 39–68, 2007.

[17] M. Raya and J.-P. Hubaux, "The security of vehicular ad hoc networks," In *Proceedings of the 3rd ACM Workshop on Security of Ad Hoc and Sensor Networks (SASN '05)*, Alexandria VA, USA, November 2005, pp. 11–21.

[18] N. W. Wang, Y. M. Huang, and W. M. Chen, "A novel secure communication scheme in vehicular ad hoc networks," *Journal Computer Communications*, vol. 31, no. 12, pp. 2827–2837, 2008.

[19] G. Samara, W. A. H. Al-Salihy, and R. Sures, "Security issues and challenges of vehicular ad hoc networks (VANET)," In *Proceedings of the 4th International Conference on New Trends in Information Science and Service Science (NISS '10)*, Gyeongju-si, Republic of Korea, May 2010, pp. 393–398.

[20] G. Calandriello, P. Papadimitratos, J.-P. Hubaux, and A. Lioy, *"Efficient and robust pseudonymous authentication in VANET,"* In *Proceedings of the Fourth ACM International Workshop on Vehicular Ad Hoc Networks - VANET '07*, Montreal Quebec, Canada, 2007, p. 19.

[21] S. Zeadally, R. Hunt, Y. S. Chen, A. Irwin, and A. Hassan, "Vehicular ad hoc networks (VANETS): Status, results, and challenges," *Telecommunication Systems*, vol. 50, pp. 217–241, 2010.

[22] T. Winter, "RPL: Ipv6 routing protocol for low-power and lossy networks," No. rfc6550. 2012. https://www.rfc-editor.org/rfc/rfc6550.html

[23] C. Cobarzan, J. Montavont, and T. Noel, "Analysis and performance evaluation of rpl under mobility," In *2014 IEEE Symposium on Computers and Communication (ISCC)*, 2014, Funchal, Portugal, pp. 1–6. IEEE.

CHAPTER 12

Edge Computing in the Digital Era

The Nexus of 5G, IoT and a Seamless Digital Future

Zahid Rasheed and Yong-Kui Ma

Harbin Institute of Technology

Inam Ullah

Gachon University

Yuning Tao

South China University of Technology

Ijaz Khan

Harbin Institute of Technology

Habib Khan

Islamia College University

Muhammad Shafiq

Guangzhou University

12.1 INTRODUCTION

The evolution of wireless communication has profoundly impacted various aspects of human life [1]. In the early days of commercial mobile services in the 1980s, mobile devices were primarily used for voice communication, with data transfer rates capped at

DOI: 10.1201/9781032648309-15

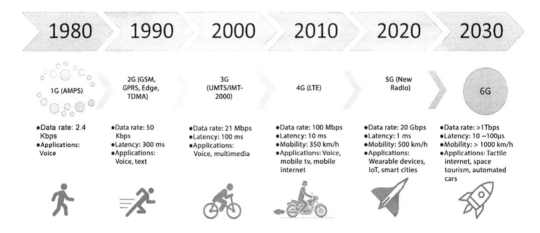

FIGURE 12.1   Wireless communication evolution. See [2].

2.4 kbps. Since then, the wireless telecommunications sector has made significant strides in speed, latency, security, and mobility. Present considerations are centered on improving data transmission speeds beyond the one terabits per second (Tbps) mark, with the fifth generation (5G) rollout already in motion and the sixth generation (6G) in its developmental phase. High data rates and superior mobility are expected to drive novel applications, including the tactile internet, fully autonomous vehicles, and holographic communication. Figure 12.1 shows the evolutionary journey of wireless technology from its first generation (1G) to the upcoming 6G, highlighting each generation's defining attributes, data rates, latency, and mobility specifics.

Edge computing, or Mobile Edge Computing or Multi-Access Edge Computing in telecommunications, provides computational and storage resources closer to end users. Edge computing can be found in various settings, including manufacturing hubs, homes, transportation systems, and personal vehicles. Service providers, including telecommunication entities, are in charge of the edge infrastructure. Diverse use cases necessitate the deployment of various applications across numerous locations. A distributed cloud, recognized for hosting applications across several sites under a unified operational framework, is particularly effective in such instances. The merits of edge technology comprise reduced latency, higher bandwidth, advanced device processing, efficient data offloading, and reliable computing and storage. Edge computing [3] is a unique computational paradigm facilitating seamless applications between the device and the extensive cloud. It notably reduces latency by shifting processing closer to users, apparatus, or data sources. Telecom edge computing refers to decentralizing data processing beyond the network edge. It also extends to the user edge, regulated by the telecommunication operator. Users can substantially reduce backhaul traffic costs and volume by deploying low-latency applications and employing near-source data processing or caching. On-premise edge computing leverages computing assets on a client's site, typically under a network operator's supervision. It is tailored to support applications and services. These processes develop within a distributed edge architecture in virtualized environments, especially concerning cloud-based activities. Employing on-premise edge computing ensures the retention of sensitive information within local infrastructure while capitalizing on the edge cloud's adaptive prowess.

The edge cloud concept blends the advantages of edge computing, a virtualized infrastructure, and commercial models. It integrates the perks of cloud computing with on-site servers, bestowing improved adaptability and scalability to efficiently cater to unexpected workload surges due to increased user activity. A private cloud is a model exclusive to select users via a private network. While it offers similar benefits to a public cloud, including scalability and flexibility, it provides enhanced security and data privacy through on-site cloud technologies. The network edge signifies the juncture where networks owned by enterprises intersect with external ones. As shown in Figure 12.2, the edge is the transitional zone between the end device and the wider internet or cloud. In this context, telco edge computing can be seen as a subset. Many locations exist to deploy telecom edge computing within public and private network sectors; cellular towers have more extensive coverage closer to end users than street cabinets. Low-latency communication systems can be vital for autonomous vehicles, allowing real-time reactions based on immediate surroundings and distant occurrences.

Telco Edge represents the combined benefits of local computing, such as on-site infrastructure and cloud computing. By employing edge computing, clients can efficiently reduce backhaul traffic costs and volume through the efficient execution of low-latency applications and data processing or caching closer to its origin. Moreover, it's crucial for telco edge computing to match the flexibility and scalability of traditional cloud computing. Telco edge computing can propel rapid organizational expansion by accommodating sudden workload surges due to unanticipated spikes in end-user traffic. This adaptability, especially in mobile applications, mandates the dynamic allocation and adjustment of computational resources across different telecom edge sites.

Edge computing, a relatively nascent domain in computing, brings cloud services and utilities closer to users. It stands out for its swift data processing and application response capabilities. Applications like surveillance, virtual reality, and real-time traffic monitoring, which demand speedy processing and response times, often operate on mobile devices with constrained resources [4,5]. The primary service and processing duties are executed on cloud servers. As depicted in Figure 12.3, mobile devices face significant challenges in latency and mobility when leveraging cloud services [6].

Edge computing provides to application demands by shifting processing capabilities closer to the network's periphery [7]. Within edge computing, three primary approaches

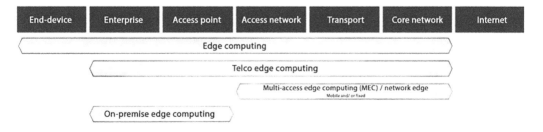

FIGURE 12.2   System composition and working principle.

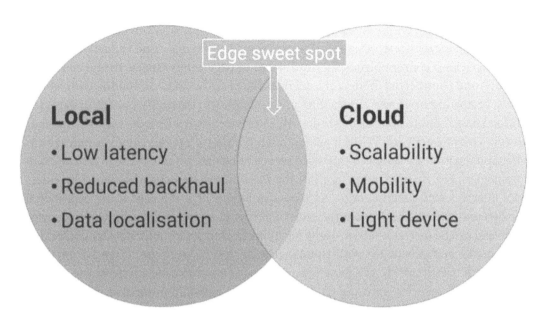

FIGURE 12.3   Illustrate the edge sweet spot.

have been highlighted as alternatives to tackle the challenges associated with traditional cloud computing: Cloudlets [8], Fog Computing [9], and Mobile Edge Computing [10]. Mobile Edge Computing, an initiative by the European Telecommunications Standards Institute (ETSI), offers computing services at base stations (BS), thereby enhancing accessibility for mobile users. Fog Computing, a concept promoted by Cisco, allows applications to operate right at the network edge through a host of smart connected devices. Cloudlets, introduced by Satya Narayanan et al., answered latency concerns associated with cloud access. They leveraged local computing resources to alleviate these latency issues. Figure 12.4 illustrates that Edge computing boasts mobility support, location awareness, ultra-low latency, and user proximity. These features make it ideal for applications ranging from industrial automation to smart maritime monitoring. Devices like routers, access points, and BS function as intermediaries, bridging the gap between intelligent mobile devices and the cloud [11].

12.2 SIGNIFICANCE OF EDGE COMPUTING

5G technology, signaled as the substance for reducing latency and boosting performance in interactive applications, underscores the pivotal role of edge computing [13]. By decentralizing and bringing cloud computing capabilities closer to user equipment (UEs), edge computing bridges the evident gap that conventional cloud computing models encounter, especially when high quality of service (QoS) standards, such as minimal latency and robust throughput, are in demand [14]. It is further accentuated by the fact that cloud servers, typically situated centrally, might increase the energy consumption of UEs due to the long distances involved [15]. In this context, the imperative for real-time packet distribution among autonomous vehicles, with an end-to-end delay below 10 ms, showcases the inefficiency of the conventional cloud approach, which stands at over 80 ms.

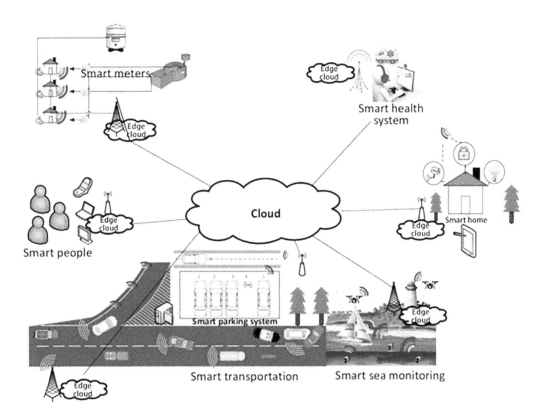

FIGURE 12.4   Edge computing application. See [12].

12.3 COMPUTING PARADIGMS

Cloud Computing: Recognized for providing users on-demand access to various services via a shared resource pool [16]. Cloud computing offers pivotal services like infrastructure as a service (IAAS), platform as a service (PAAS), and software as a service (SAAS) [17]. It optimizes shared resources across users, tailoring them to their time zones, ensuring efficient allocation and utility.

Edge Computing is characterized by its relocation of computational data, services, and applications from centralized servers to the network's periphery [18]. Edge computing offers unmatched bandwidth, minimal latency, and real-time network data access [19]. This decentralized approach is geared toward enterprises and end users, showcasing potential in services like location assistance, augmented reality, and data caching. The crux of its distinction from cloud computing lies in its geographical deployment and inherent mobility support.

Comparison between Cloud and Edge Computing: The core distinction revolves around their geographical deployment, as shown in Figure 12.5. Cloud computing is majorly internet-hosted; edge computing operates at the network's periphery. It leads to lower latency in edge computing owing to reduced proximity between mobile devices and servers [20]. Moreover, edge computing, tailored for edge users, inherently provides location awareness and a decentralized server deployment strategy, reducing the chances of data in route

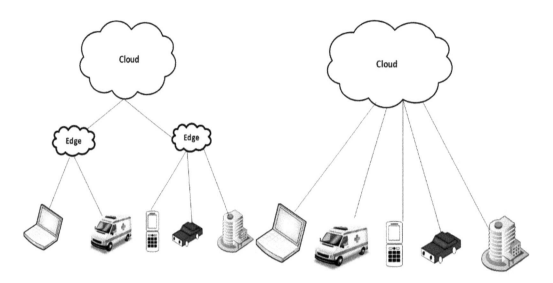

FIGURE 12.5 Graphically represents the major distinctions between edge and cloud computing. See [21].

attacks. Despite their myriad benefits, edge computing does face scalability limitations due to its hardware constraints.

Fog Computing: Introduced by Cisco in 2012, this model decentralizes cloud computing infrastructure, transferring nodes closer to the network's periphery, thereby better managing applications and addressing high-speed internet demands [22]. It lessens the network load, alleviates traffic delays, and simplifies network and device management. One can discern a difference between fog computing and MEC. At the same time, the former involves resource relocation from the central cloud to the network's edge; MEC implements fog computing with computational resources moved closest to the network's edge, usually at BS [23].

Moisture Computing: An innovative distributed computing paradigm introduced around 2021, moisture computing marries the advantages of cloud and mobile edge computing [24]. It positions its processing and storage infrastructure closer to the end user, offering a middle ground between MEC and traditional cloud computing. Table 12.1 summarizes the distributed computing paradigms.

12.4 EDGE COMPUTING AND SIMILAR CONCEPTS

Edge computing is an expansion of cloud computing in which computational services are brought closer to the end users at the network's edge. The edge vision was developed to resolve the issue of excessive latency in delay-sensitive services and applications poorly served by the cloud computing paradigm. The following requirements must be met by these applications: (a) extremely low and predictable latency, (b) location awareness, and (c) mobility support. Even though edge computing has numerous advantages over cloud computing, this field of study is still in its infancy. Edge computing is a self-contained

TABLE 12.1   Summary of Computing Paradigms

	Fog Computing	Mobile Edge Computing	Could Computing	Moisture Computing
Year of introduced	2012	2014	2009	2021
Supporting organization	OpenFog Consortium	ETSI	Open Edge Computing initiative	None
Driving force	Internet of Things and smart cities application	5G requirements and integration	Mobile applications in community places	5G, Internet of Things, and smart cities applications
Support features	OpenFog reference architectures	ETSI MEC architecture	Open stack ++ platform	Moisture computing architecture
Computing infrastructure location	Anywhere between cloud and end devices	Edge of network base station	Edge of the network (typically one hope away from the end devices)	Above the edge of the network (typically two hops away from the end devices)

computer paradigm composed of dispersed heterogeneous devices that interface with the network and execute computing functions, including storage and processing [25].

These responsibilities may also contribute to the delivery of lease-based services, in which a user leases a device in exchange for incentives. According to Cisco, fog computing is an extension of cloud computing that transfers resources and services to the periphery network. The edge network is a virtualized platform with storage, computation, and networking capabilities. Cloudlet and mobile edge computing [12] are concepts comparable to the Fog computing paradigm. The purpose of cloudlet and mobile edge computing is to provide services to mobile users while enabling them to utilize locally accessible resources. In contrast, fog relies on Cisco-designed equipment with computational, router, and switch capabilities.

12.5 EDGE COMPUTING CHARACTERISTICS

While sharing commonalities with cloud computing, such as data processing and storage capabilities, edge computing exhibits unique characteristics that set it apart. These attributes make edge computing suitable for real-time applications and localized data processing. The following elements are noteworthy.

12.5.1 Dense Geographical Distribution

Edge computing is a paradigm that involves the deployment of many computing platforms on edge networks, hence facilitating the proximity of cloud services to end users [26]. The broad geographical dispersion of infrastructure has several benefits: (a) Network administrators can offer location-based mobility services within the WAN without the need to traverse its entirety; (b) The utilization of big data analytics allows for faster and more precise analysis, as shown by previous research [27]; (c) Edge systems facilitate the execution of large-scale real-time analytics. This capability has been acknowledged in the literature [28].

12.5.2 Mobility Support

Edge computing enables enhanced mobility through protocols like the Locator ID Separation Protocol (LISP), which establishes direct connections with mobile devices in response to the rapid proliferation of such devices. The LISP protocol facilitates establishing a distributed directory system by separating location identification from host identity. The fundamental concept underlying mobility support in edge computing involves decoupling the host identity from the location identity.

12.5.3 Location Awareness

The location-awareness feature of edge computing enables mobile users to access services from the edge server close to their current location. Individuals can employ diverse technological tools to locate electronic devices, such as cell phone infrastructure, GPS systems, and wireless access points. Several edge-computing applications can utilize this geographical awareness, including fog-based vehicle safety and edge-based catastrophe management.

12.5.4 Proximity

Edge computing involves the relocation of processing resources and services in closer proximity to end users, hence enhancing their overall user experience. Users can employ network context information to make offloading and service consumption decisions. These decisions are dependent on the presence of computing resources and services within their local vicinity. Similarly, the service provider can leverage the knowledge of mobile users to enhance their services and optimize resource allocation. It can be achieved through the extraction of device-specific data and the evaluation of user behavior.

12.5.5 Low Latency

The notion of edge computing involves the relocation of processing resources and services in closer proximity to end users, resulting in a reduction in latency during service access. The low latency of edge computing enables users to execute resource-intensive and delay-sensitive applications on edge devices with ample resources, such as routers, access points, BS, or dedicated servers.

12.5.6 Context-Awareness

Context awareness is a characteristic shown by mobile devices, which can be defined in conjunction with location awareness. Using context information from the mobile device can inform decision-making regarding offloading choices and accessing edge services in edge computing [29]. Context-aware services can be delivered to users at the edge by leveraging real-time network data, including information on network load and user location. In addition, the service provider can utilize contextual information to enhance customer satisfaction and improve the overall quality of the customer experience.

12.5.7 Heterogeneity

Edge computing systems utilize many platforms, topologies, infrastructures, processing mechanisms, and communication technologies, together known as heterogeneity. This heterogeneity encompasses a range of components such as end devices, edge servers, and networks. Several primary factors contribute to heterogeneity in end devices, namely software, hardware, and technical variations. Various factors, including the utilization of APIs, the implementation of custom-built rules, and the deployment of different platforms, influence the heterogeneity of edge servers. These variations give rise to obstacles in achieving interoperability, posing a significant undertaking in successfully deploying edge computing. The concept of "network heterogeneity" pertains to the diverse array of communication technologies that impact the provision of edge services.

12.6 KEY REQUIREMENTS OF EDGE COMPUTING IN 5G

There are several key requirements for the efficient implementation and functioning of edge computing within the context of 5G. Considering a balanced trade-off is vital when evaluating the significance of all key needs, as it varies based on the specific applications.

The key incentive for utilizing edge computing instead of cloud computing is the ability to engage in real-time engagement. The primary objective is to guarantee minimal latency to accommodate time-sensitive applications and services such as remote surgery, tactile internet, ultra-reliable low-latency communication (URLLC), crewless vehicles [30], and car accident avoidance. Additionally, this aim is to enhance the quality of service. Edge servers can offer diverse real-time services, encompassing decision-making processes and data processing activities. Furthermore, the feasibility of local processing arises from the capability to handle data and user requests locally instead of relying on cloud-based edge servers. It implies that two significant outcomes can be achieved by reducing the volume of traffic transmitted between a small cell and the core network: (1) enhancement of the connection's bandwidth to prevent congestion and (2) reduction of the overall traffic load inside the core network. Besides, it is important to note that edge clouds necessitate a substantial data rate to effectively transmit the vast quantities of data generated by diverse applications such as virtual reality and remote surgery [31]. Edge servers, which may be seamlessly incorporated into BS, provide a convenient means of establishing connections with edge clouds without the need to traverse the primary network. Utilizing millimeter wave (mm-wave) frequency bands within a confined cellular environment facilitates data transfer at elevated rates. Additionally, it is important to note that high availability plays a crucial role in ensuring the accessibility of cloud services at the edge. The availability of edge clouds is paramount because edge computing involves transmitting data and application logic to these cloud systems.

12.7 MULTI-ACCESS EDGE COMPUTING (MEC)

With the advent of 5G technology, academics have focused on challenges associated with applications that necessitate substantial computational power, storage capacity, and real-time data processing. The diligent endeavors led to establishing and refining mobile edge computing concepts, whereby computational tasks were shifted toward the network's periphery. MEC has led to the development of various technologies, including fog computing, cloudlet computing, and mobile cloud computing. Nevertheless, each approach has its distinct limitations. For instance, fog computing depends on cloud computing data centers and cannot function autonomously with its own managed data center. The integration of this system into the mobile network is limited, and it frequently works independently [32].

In contrast, the telecoms industry exhibited a significant interest in MEC. The MEC-ISG initiative was introduced by the ETSI in 2014 to promote the concept of mobile edge computing. The authors wanted to integrate MEC into the CRAN architecture [33]. The terminology of mobile edge computing was altered to multi-access edge computing, a term that specifically pertains to the C-RAN 5G architecture. The objective of this Mobile Edge Computing Industry Specification Group (MEC ISG) was to establish a standardized framework for mobile edge computing in the context of 5G technology. The focus of their efforts was directed towards enhancing network optimization and user service efficiency by bringing computer capabilities in closer proximity to the end user [34]. The intricacy of the framework is depicted in Figure 12.6, which provides a visual depiction that clarifies the numerous interactions among different components within the MEC ecosystem. This diagram offers a comprehensive visual understanding of the functional structure of the framework. The mobile edge host level, a crucial component of the MEC infrastructure, encompasses the mobile edge host and its related management systems. These systems are responsible for providing the necessary virtualized architecture and platform required

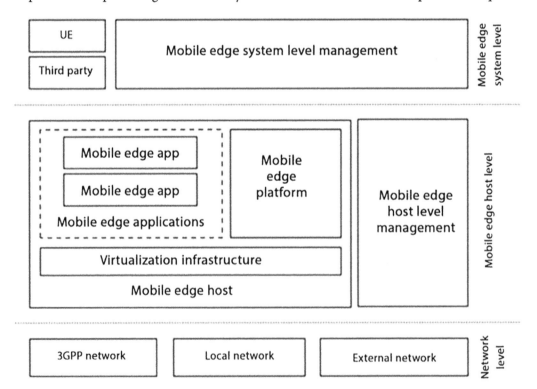

FIGURE 12.6   ETSI MEC framework. See [35].

for mobile edge applications. The network level complements the core by facilitating connectivity across various networks, including 3GPP mobile networks and the Internet. The structure, overseen by the mobile edge system level management, is designed to abstract the complexities of lower-level operations. This abstraction enhances user and third-party access, optimizing interactions within the MEC framework.

Several industry initiatives have been implemented to establish standardization for MEC and its associated technologies in the context of 5G. One such project is the Open Fog Consortium, established in 2015 to develop a unified open framework for distributed computing. Furthermore, the development of 5G MEC has led to setting 3GPP standards [36].

12.8 MEC AND 5G INTEGRATION

A comprehensive study was done by scholars and professionals from academia and business to understand better the various aspects involved in integrating 5G with MEC and the challenges and opportunities associated with this integration. This section is dedicated to exploring the integration of 5G with MEC, aiming to analyze MEC's role in 5G and subsequent generations of wireless networks. The study encompasses a comprehensive analysis of the fundamental aspects of MEC, the integration of MEC with existing technologies, and the state-of-the-art [37] in MEC implementation [38–41]. The subject of art is a broad and multifaceted field that encompasses various forms of creativity. MEC and its associated technologies proposed that integrating 5G with MEC would be beneficial. The researchers analyzed the increase in data traffic after deploying 5G technology. The escalation in data traffic would impose significant pressure on the 5G infrastructure, posing challenges in the operation of applications that necessitate higher computational capabilities and lower latency thresholds. However, the attention does not extend to applications within the smart city context with distinct requirements. The significance of the location where the computation is performed is examined in a study by authors in reference [42]. The authors engaged in a discourse on cloud architecture, analyzing the advantages and disadvantages of its implementation. They discussed the MEC architecture and its potential benefits within a 5G environment. The study revolved around Software Defined Networking (SDN) and Network Function Virtualization (NFV) and their interrelation with MEC. In addition to this, a study [35] examined the revolutionary effects of MEC on the telecommunications industry, offering a comprehensive comparative analysis of MEC concerning traditional cloud computing infrastructures. This line of investigation pertains to the orchestration elements of MEC, specifically focusing on the operational platforms and application scenarios involved.

Furthermore, the authors discussed the collaborative endeavors undertaken by standards bodies to establish a cohesive framework for implementing and administrating MEC. The research discussed in reference [43] highlights the significant importance of MEC in the infrastructure of the 5G radio access network. It specifically emphasizes the potential improvements in latency that might be achieved through implementing MEC. This topic is further expounded upon in the cited reference [44], which comprehensively examines three distinct 5G use cases and assesses the potential impact of MEC on each scenario. The report acknowledged the difficulties associated with incorporating MEC and 5G technologies, emphasizing the necessity for ongoing innovation and resolution of issues during this integration procedure. The research discussed in citation [45] examined the relationship between MEC, 5G, and the Internet of Things (IoT), shedding light on the crucial technologies that facilitate the efficient utilization of MEC within this interconnected framework. The topics covered in this study encompass an examination of cloud computing, SDN/NFV, virtual machines (VMs), and intelligent devices, all of which have the potential to benefit from the implementation of MEC. In summary, the authors presented a comprehensive assessment of the role of MEC in the context of the 5G and Internet of Things (IoT) frameworks. They delved into several approaches for integrating MEC into these paradigms while highlighting the ongoing obstacles and unresolved matters that define the convergence of MEC, 5G, and IoT.

12.9 IoT AND EDGE COMPUTING ARCHITECTURE

This section primarily centers on the architectural aspects of edge computing and the utilization of IoT devices and technologies predicated on edge computing principles. Although there are variations in the capabilities and services these devices offer, certain essential aspects distinguish them. The services offered include complex event processing, artificial intelligence models, offline support, data administration, and various applications. One of the fundamental characteristics that should be incorporated into the next generation of IoT edge devices is the ability to swiftly modify device configurations using remote tools alongside enhanced security measures for packet transmission and software updates. Figure 12.7 illustrates the architectural framework of edge computing, showcasing its hierarchical structure, including multiple levels. The initial level within this framework pertains to IoT devices [46,47]. Resource pooling techniques can facilitate remote communication, which involves aggregating resources. These resources are subsequently transmitted to edge servers, such as micro clouds and cloudlets, as well as some intelligent routers and switches. It enables further data processing to take place. Alternatively, preliminary data preprocessing may be conducted in certain instances before the ultimate cloud processing stage. The second tier of this architectural framework is represented by the edge server platform, which serves as a gateway for accessing network resources such as storage and computational capabilities. The facility's proximity to a high-demand IoT client reduces latency and improves load processing speed. The infrastructure-based cloud is the third tier of cloud computing, serving as the primary platform for data processing and computational tasks. The data generated by IoT devices and sensors is digitally recorded and may be accessed by any device at various levels through web platforms.

12.9.1 Characteristics of IoT in Edge Computing

The IoT encompasses a network of interconnected devices communicating via the Internet, including sensors, telephones, wearable gadgets, and machines. The term "Internet of Things" (IoT) encompasses machine-readable devices capable of digitizing data. Advantages of edge computing include reduced latency, enhanced location awareness, extensive geographical coverage, support for mobility, and a large number of interconnected nodes. Edge computing facilitates the independent operation of enterprises inside a certain area,

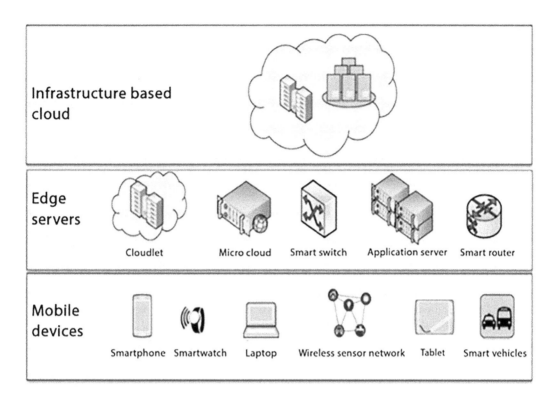

FIGURE 12.7 The mobile edge cloud computing (MECC) architecture. See [48].

region, or domain while adhering to local security limitations [48]. It is achieved by leveraging public or private cloud resources and local computing capabilities. Edge computing refers to the practice of doing data processing close to the network's edge. The architecture of edge computing is characterized by its distributed nature and adherence to an open information technology framework. The scope of the system encompasses a wide range of components, spanning from end users through the edge, core, and cloud [49–52].

12.10 SMART CITIES AND EDGE COMPUTING

The notion of "Smart Cities" is currently prevalent globally. This part comprehensively analyzes the topic, focusing on its relevance and practical implications for the human population. The concept of the smart city can be conceptualized as a framework. This "framework" construction primarily relies on state-of-the-art technology, particularly information and communication. The objective of formulating and executing this plan is to manage the pace of urbanization effectively. Governments are facing growing challenges in administrating large communities and providing essential services to residents. The predominant technology in smart cities is wireless and seamless, primarily facilitating communication without direct human intervention. Data is frequently gathered in the cloud through many sensors and gadgets for research and decision-making. The smart city ecosystem encompasses various stakeholders, including citizens, government entities, corporations, municipalities, and other relevant actors. Implementing multiple IoT devices, platforms, and frameworks in a smart city can reduce energy consumption, efficient traffic management during peak hours, and timely provision of emergency services. Cities can undergo transformations that result in enhanced cleanliness and environmental sustainability.

Smart cities can be categorized into three distinct layers or tiers. At a fundamental level, it is possible to analyze the many technological devices employed for data collection that are readily accessible. Mobile phones and sensors exemplify these technological devices. Generally, these devices are connected to a highly efficient communication network, such as the 5G network. All programs on these devices can be categorized at the second level. In machine-to-machine (M2M) communication, these applications facilitate data transformation into meaningful information for end users or other interconnected devices. At this stage, the stakeholders involved in the application development and the companies responsible for providing the necessary technology are identified. The third level of analysis focuses on consumers' actual utilization of the programs. Individuals and inhabitants are unlikely to avail themselves of services unless they are deemed advantageous and feasible.

The issue of privacy and security breaches represents a significant concern within the context of smart cities. The enhancement of urban security can be achieved by implementing various intelligent devices, such as digital cameras, intelligent transportation systems, and people monitoring systems designed to ensure the safety of citizens. Nevertheless, it is important to acknowledge that all of these methods possess the capacity to infringe upon individuals' privacy. Consequently, achieving a harmonious equilibrium between these two constituent elements is imperative. Numerous research endeavors are underway to support attaining diverse objectives associated with smart city initiatives. In reference [43], the authors aim to enhance the user experience of video streaming applications within the context of smart cities. They proposed the utilization of MEC technology as a means to improve the quality of service delivery. The main focus areas revolved around mobility, latency, and network congestion. Establishing a MEC service is recommended to offer users expedited accessibility. Nevertheless, the writers do not address using 5G technology to accomplish their goals.

Likewise, the authors of reference [53] provided an analysis and discourse on the development and functionality of 5G technology. This article examined the security ramifications of 5G technology and its impact on existing technologies such as the IoT, autonomous vehicles, and artificial intelligence (AI). Once the challenges of implementing the 5G network are effectively addressed, it can serve as a fundamental framework for integrating intelligent urban applications encompassing several domains, including transportation, public services, healthcare, and infrastructure. The scholarly article [54] examined the impact and ramifications of 5G technology on Intelligent Transportation Systems (ITS), among the several applications within smart cities. The article discussed the technological dimensions of the economic benefits and sectors that might be influenced by smart cities, including but not limited to public transit, manufacturing, health care, and energy. In many urban centers across the globe, many endeavors have been initiated and continue to be in progress. The provided data presents a compilation of several smart cities across the world. Based on the rankings provided by Forbes, the ten leading smart cities globally are London, New York, Paris, Tokyo, Reykjavik, Copenhagen, Berlin, Amsterdam, Singapore, and Hong Kong, arranged in ascending order from one to ten. Based on the findings of Cities in Motion, the United States' top ten smart cities are ranked as follows: New York, Los Angeles, Chicago, San Francisco, Washington, Boston, Miami, Phoenix, Dallas, and San Diego, respectively, in descending order from first to tenth. Based on a report from the computer world, the United Kingdom's leading smart cities are ranked as follows: Milton Keynes, Glasgow, Nottingham, Cambridge, Bristol, London, Manchester, Birmingham, and Leeds, in ascending order from first to tenth place.

Using wireless sensor networks [55], digital networks, and apps contributes to advancing technology in smart cities by integrating the IoT and facilitating the Internet of Vehicles (IoV). The objective of converting a city into an intelligent environment is to address the challenges posed by urbanization and population expansion. A smart city is an urban area that effectively attains sustainable economic development while upholding a superior quality of life. Several technologies have been identified as potential contributors to sustainable urbanization. These include intelligent solutions to mitigate transportation congestion, implementing environmentally friendly structures, and utilizing advanced industrial control systems (ICS). A smart city encompasses the strategic utilization of technology to optimize various aspects of human life, including living conditions, work environments, commuting systems, and information exchange. Next-generation autos, integral to the broader concept of the IoT and include advanced sensing, communication, and social functionalities, represent a significant component of smart cities [56–61]. Vehicles have the potential to contribute to the realization of smart cities through their capacity to provide mobile wireless sensing and communication capabilities. Intelligent automobiles will establish communication links with navigation, broadcast satellites, passenger cell phones, roadside gadgets, and other intelligent vehicles. This integration positions them as a pivotal element within the IoT framework and the advancement of smart city infrastructure [62].

12.11 EDGE COMPUTING: ADVANTAGES AND DISADVANTAGES

1. **Advantages of Edge Computing:**

   - **Reduced Latency:** Edge computing augments network efficiency by minimizing latency, a common issue in traditional networks that often leads to delay, including in proximal communications. By processing data at the network's periphery, edge computing substantially mitigates such delay, thereby accelerating the speed of data exchange.

   - **Traffic Optimization:** Edge centers are designed to bypass centralized systems bottleneck challenges, enhancing overall network efficiency.

   - **Enhanced Security:** Edge computing decentralizes processing and storage capabilities. This approach diminishes vulnerabilities like distributed denial-of-service (DDoS) attacks. If a singular machine is compromised, the perpetrator's access is confined to that specific computer's data.

- **Cost-Effective Scalability:** Integrating IoT devices with edge servers augments operational capabilities without constructing expensive data centers. It also minimizes bandwidth strain, allowing for more efficient expansions [56].

- **Flexibility and Speed in Market Expansion:** Companies can swiftly deploy edge devices for service provision. If the market proves unsuitable, retracting the services remains cost-effective.

- **Consistent User Experience:** Edge servers' proximity to end users minimizes the risk of network outages, ensuring consistent, uninterrupted services.

2. **Disadvantages of Edge Computing:**

- **Limited Resources:** Edge devices, owing to their decentralized nature, often possess fewer computational resources compared to centralized cloud servers, which might lead to reduced processing speeds.

- **Increased Security Vulnerabilities:** The decentralized edge computing approach could expand the attack surface for cybercriminals. Ensuring robust security in such a distributed setting poses challenges.

- **Financial Considerations:** While edge computing can offer cost savings in certain scenarios, setting up the necessary infrastructure without a local edge partner can be financially intensive. Maintenance of multiple devices across varied locations can also inflate costs.

- **Obsolescence Risk:** The rapidly evolving technological landscape means devices and software can become outdated swiftly, necessitating frequent upgrades or replacements.

12.12 INTEGRATING EDGE COMPUTING IN THE SPHERE OF 5G: DIVERSE APPLICATIONS AND THEIR IMPLICATIONS

The interplay between 5G and edge computing promises to redefine various industries by ensuring real-time engagement, local processing, superior data throughput, and consistent availability.

- **Healthcare:** This sector is set to experience transformative changes. Beyond facilitating telemedicine, edge computing allows for real-time remote surgeries and patient diagnostics. Medical professionals can leverage remote platforms to operate surgical equipment precisely, ensuring patient safety. Additionally, real-time patient vital signs monitoring becomes seamless, advancing proactive healthcare.

- **Entertainment and Multimedia:** With the bandwidth capabilities of 5G, streaming high-definition content like HDTV and 3D TV becomes efficient. Additionally, immersive experiences through virtual reality (VR), augmented reality (AR), and mixed reality (MR) are enhanced. By outsourcing the processing tasks from VR glasses to edge servers, the hardware can potentially be made more compact and user-friendly.

- **Tactile Internet:** This innovation heralds a new chapter in the Internet of Things. The tactile internet enables the instantaneous transmission of control signals and tangible sensations over vast distances. It promises an exceptionally responsive and stable network connectivity, expanding IoT capabilities.

- **URLLC (Ultra Reliable Low Latency Communications):** Ensuring robust reliability, especially in Machine-to-Machine (M2M) communication, URLLC focuses on transmitting short data packets with minimal delay. It ensures high reliability, critical for applications like fire alarms and emergency response systems.

- **IoT:** Many intelligent devices, from household appliances to industrial sensors, connect to the internet, streamlining processes and enhancing user experience. With edge computing, these devices can process data locally, reducing latency and improving responsiveness.

- **Advanced Manufacturing:** The future of manufacturing lies in automation and precision. Edge computing facilitates remote operation of heavy machinery, especially in hazardous or hard-to-reach areas, ensuring both efficiency and operator safety.

- **Emergency Response:** In critical situations, swift decision-making is paramount. Edge computing aids in collecting diverse data in real-time, ensuring emergency response teams, such as firefighters, receive real-time insights before reaching an incident scene. It enables a more proactive and informed approach to crisis management.

- **Intelligent Transportation System (ITS):** ITS is transforming road safety by providing drivers with real-time data from traffic centers; potential hazards can be preemptively addressed. Furthermore, with the advent of autonomous vehicles, edge computing plays a pivotal role in real-time data processing, ensuring they navigate their environment safely [53].

12.13 DISCUSSION

The modern digital landscape is experiencing a transformative shift as the contours of data processing evolve. A closer examination of edge computing, set against the backdrop of innovations like 5G, IoT, and the visionary concept of smart cities, reveals deep intricacies and vast potentialities. Historically, dependence on centralized cloud configurations dominated the approach to data processing. Yet, the journey from these traditional setups through intermediary stages like fog and moisture computing has culminated in the rise of edge computing. This decentralized approach is a testament to the ongoing metamorphosis in managing and processing data. One cannot overlook its significant impact on latency and overall network performance. Fields such as healthcare, where the milliseconds matter—for instance, in remote surgeries—benefit immensely from this reduction in latency. But this revolution is multi-dimensional; the emergence and growth of 5G technology, characterized by its high-speed data transfers, promises to catapult the advantages of edge computing to newer zeniths. Especially when considering the interaction of MEC with 5G, potentially looking at a seamless digital realm with minimized delays and an exponentially enhanced user experience. Yet, in every transformation lies inherent challenges. The distributed nature of edge computing, while safeguarding against large-scale centralized attacks, brings forth a spectrum of security concerns. Each edge device, potentially devoid of centralized monitoring, becomes vulnerable if not updated and secured. However, the silver lining might be found in the inherent nature of 5G, which could facilitate rapid security updates, creating a dynamic shield against potential threats.

The strategic implications of edge computing span beyond just security. Organizations grapple with the logistical challenges of establishing and maintaining this expansive network. The economic repercussions also demand attention, weighing the undeniable benefits against the fiscal demands of such an extensive, decentralized model. IoT's explosion further accentuates the role of edge computing. With countless devices transmitting data, the need for real-time, localized processing has never been more pronounced. This interdependence forms the bedrock for future smart cities, wherein the symbiotic relationship will be crucial for operations ranging from intelligent traffic systems to energy-efficient smart grids. While lauding the strides edge computing has taken, it's imperative to maintain a balanced perspective. Its myriad advantages, from latency reductions to user experience enhancements, are juxtaposed against its challenges. Yet, as sectors from healthcare to transportation stand to be revolutionized by this technological marvel, the future is undeniably promising. As the 5G era dawns upon us, the convergence of this high-speed technology with edge computing signals a paradigm shift in our digital narrative. It's a future laden with potential, but it calls for industry-wide collaboration, persistent research, and significant investment to unlock this digital amalgamation's wonders.

12.14 CONCLUSION

In our comprehensive exploration of the contemporary digital realm, edge computing emerges as a concept and a transformative force driving the next wave of technological advancement. Its significance is underscored by its departure from traditional computing paradigms, leading to a decentralized model that brings computation closer to data sources. This shift promotes enhanced responsiveness, marked by characteristics like reduced latency, robust mobility support, and improved geographical distribution. The advent of MEC elevates this proposition further. MEC optimizes network resources as a bridge between edge devices and centralized systems, and its integration with 5G technology crystallizes this. The fusion promises ultra-fast data transmission and a robust infrastructure that supports the burgeoning demands of modern applications. The IoT and edge computing are symbiotic. With billions of interconnected devices generating vast amounts of data, edge computing provides the perfect platform for timely data processing and decision-making right at the source. It is particularly salient in smart cities, where real-time data can be harnessed for many functions—traffic management and energy conservation. Yet, as with any significant technological evolution, edge computing also brings challenges. Given the dispersed nature of devices, infrastructure overheads, and the financial nuances of deploying such a widespread network, these encompass security vulnerabilities. However, as seen in varied applications, the advantages overshadow these challenges. Healthcare, transportation, entertainment, and manufacturing sectors stand as testament. Envision remote surgeries being executed flawlessly thanks to negligible latency or smart transportation systems that can predict and adapt to real-time changes. In essence, the confluence of edge computing with other technological marvels like 5G and IoT signals a promising horizon for digital advancements. As industries and policymakers march ahead, it will be paramount to remain vigilant, ensuring that innovation is balanced with security and inclusivity. The journey ahead is rife with potential. By leveraging these technologies judiciously and investing in continual research and infrastructure, a future that is not only digital but also efficient, secure, and encompassing awaits.

REFERENCES

[1] I. Haq et al., "Impact of 3G and 4G technology performance on customer satisfaction in the telecommunication industry," *Electronics*, vol. 12, no. 7, pp. 1–24, 2023, doi:10.3390/electronics12071697.

[2] A. Tufail, A. Namoun, A. Alrehaili, and A. Ali, "A survey on 5G enabled multi-access edge computing for smart cities: Issues and future prospects," *IJCSNS Int. J. Comput. Sci. Netw. Secur.*, vol. 21, no. 6, pp. 107–118, 2021, doi:10.22937/IJCSNS.2021.21.6.15.

[3] I. Ullah, S. Qian, Z. Deng, and J. H. Lee, "Extended Kalman Filter-based localization algorithm by edge computing in wireless sensor networks," *Digit. Commun. Networks*, vol. 7, no. 2, pp. 187–195, 2021, doi:10.1016/j.dcan.2020.08.002.

[4] N. U. Hassan, B. Tunaboylu, and M. Yasin, "Development and technological steps for realization of AB-PEM Fuel cell-a step towards clean & sustainable energy solutions," In 4th International Conference on *Power Generation Systems and Renewable Energy Technologies, PGSRET 2018*, Islamabad, Pakistan, 2019, pp. 0–5. doi:10.1109/PGSRET.2018.8685934.

[5] M. Liu, F. R. Yu, Y. Teng, V. C. M. Leung, and M. Song, "Distributed resource allocation in blockchain-based video streaming systems with mobile edge computing," *IEEE Trans. Wirel. Commun.*, vol. 18, no. 1, pp. 695–708, 2019, doi:10.1109/TWC.2018.2885266.

[6] P. Pace, G. Aloi, R. Gravina, G. Caliciuri, G. Fortino, and A. Liotta, "An edge-based architecture to support efficient applications for healthcare industry 4.0," *IEEE Trans. Ind. Informatics*, vol. 15, no. 1, pp. 481–489, 2019, doi:10.1109/TII.2018.2843169.

[7] U. Shaukat, E. Ahmed, Z. Anwar, and F. Xia, "Cloudlet deployment in local wireless networks: Motivation, architectures, applications, and open challenges," *J. Netw. Comp. Appl.*, vol. 62, no. January. pp. 18–40, 2016. doi:10.1016/j.jnca.2015.11.009.

[8] M. Satyanarayanan, P. Bahl, R. Cáceres, and N. Davies, "The case for VM-based cloudlets in mobile computing," *IEEE Pervasive Comput.*, vol. 8, no. 4, pp. 14–23, 2009, doi:10.1109/MPRV.2009.82.

[9] W. Bao et al., "Follow me fog: Toward seamless handover timing schemes in a fog computing environment," *IEEE Commun. Mag.*, vol. 55, no. 11, pp. 72–78, 2017, doi:10.1109/MCOM.2017.1700363.

[10] E. Ahmed and M. H. Rehmani, "Mobile edge computing: Opportunities, solutions, and challenges," *Futur. Gener. Comput. Syst.*, vol. 70, pp. 59–63, 2017, doi:10.1016/j.future.2016.09.015.

[11] Z. Zhang, W. Zhang, and F. H. Tseng, "Satellite mobile edge computing: Improving QoS of high-speed satellite-terrestrial networks using edge computing techniques," *IEEE Netw.*, vol. 33, no. 1, pp. 70–76, 2019, doi:10.1109/MNET.2018.1800172.

[12] W. Z. Khan, E. Ahmed, S. Hakak, I. Yaqoob, and A. Ahmed, "Edge computing: A survey," *Futur. Gener. Comput. Syst.*, vol. 97, no. February, pp. 219–235, 2019, doi:10.1016/j.future.2019.02.050.

[13] Z. Ning, X. Kong, F. Xia, W. Hou, and X. Wang, "Green and sustainable cloud of things: Enabling collaborative edge computing," *IEEE Commun. Mag.*, vol. 57, no. 1, pp. 72–78, 2019, doi:10.1109/MCOM.2018.1700895.

[14] K. Yeow, A. Gani, R. W. Ahmad, J. J. P. C. Rodrigues, and K. Ko, "Decentralized consensus for edge-centric Internet of Things: A review, taxonomy, and research issues," *IEEE Access*, vol. 6, pp. 1513–1524, 2018, doi:10.1109/ACCESS.2017.2779263.

[15] W. Hu et al., "Quantifying the impact of edge computing on mobile applications," In *Proceedings of the 7th ACM SIGOPS Asia-Pacific Workshop on Systems, APSys 2016*, Hong Kong, 2016, March. doi:10.1145/2967360.2967369.

[16] E. Simmon, "Evaluation of cloud computing services based on NIST 800-145," *Natl. Inst. Stand. Technol.*, vol. 500, no. 322, pp. 1–24, 2018.

[17] I. Yaqoob, E. Ahmed, A. Gani, A., S. Mokhtar, and M. Imran, "Heterogeneity-aware task allocation in mobile ad hoc cloud," *IEEE Access*, vol. 5, pp. 1779–1795, 2017.

[18] P. Wang, C. Yao, Z. Zheng, G. Sun, and L. Song, "Joint task assignment, transmission, and computing resource allocation in multilayer mobile edge computing systems," *IEEE Internet Things J.*, vol. 6, no. 2, pp. 2872–2884, 2019, doi:10.1109/JIOT.2018.2876198.

[19] Y. Sahni, J. Cao, and L. Yang, "Data-aware task allocation for achieving low latency in collaborative edge computing," *IEEE Internet Things J.*, vol. 6, no. 2, pp. 3512–3524, 2019, doi:10.1109/JIOT.2018.2886757.

[20] W. Li, Z. Chen, X. Gao, W. Liu, J. Wang, and S. Member, "Multimodel framework for indoor localization under mobile edge computing environment," *IEEE Internet Things J.*, vol. 6, no. 3, pp. 4844–4853, 2019.

[21] N. Hassan, K. L. A. Yau, and C. Wu, "Edge computing in 5G: A review," *IEEE Access*, vol. 7, pp. 127276–127289, 2019, doi:10.1109/ACCESS.2019.2938534.

[22] Y. Liu, M. Peng, G. Shou, Y. Chen, and S. Chen, "Toward edge intelligence: Multiaccess edge computing for 5G and Internet of Things," *IEEE Internet Things J.*, vol. 7, no. 8, pp. 6722–6747, 2020, doi:10.1109/JIOT.2020.3004500.

[23] M. Aazam, S. Zeadally, and K. A. Harras, "Fog computing architecture, evaluation, and future research directions," *IEEE Commun. Mag.*, vol. 56, no. 5, pp. 46–52, 2018, doi:10.1109/MCOM.2018.1700707.

[24] K. Dolui and S. K. Datta, "Comparison of edge computing implementations: Fog computing, cloudlet and mobile edge computing," *GIoTS 2017- Glob. Internet Things Summit, Proc.*, Geneva, Switzerland, June 2017, doi:10.1109/GIOTS.2017.8016213.

[25] L. M. Vaquero and L. Rodero-Merino, "Finding your way in the fog: Towards a comprehensive definition of fog computing," *Comp. Commun. Rev.*, vol. 44, no. 5, pp. 27–32, 2014. doi:10.1145/2677046.2677052.

[26] M. Satyanarayanan, "How we created edge computing," *Nat. Electron.*, vol. 2, no. 1, p. 42, 2019, doi:10.1038/s41928-018-0194-x.

[27] E. Ahmed et al., "The role of big data analytics in Internet of Things," *Comput. Networks*, vol. 129, pp. 459–471, 2017, doi:10.1016/j.comnet.2017.06.013.

[28] S. Josilo and G. Dan, "Selfish decentralized computation offloading for mobile cloud computing in dense wireless networks," *IEEE Trans. Mob. Comput.*, vol. 18, no. 1, pp. 207–220, 2019, doi:10.1109/TMC.2018.2829874.

[29] B. Han, S. Wong, C. Mannweiler, M. R. Crippa, and H. D. Schotten, "Context-awareness enhances 5G multi-access edge computing reliability," *IEEE Access*, vol. 7, pp. 21290–21299, 2019, doi:10.1109/ACCESS.2019.2898316.

[30] M. Fahad et al., "Grey wolf optimization based clustering algorithm for vehicular ad-hoc networks," *Comput. Electr. Eng.*, vol. 70, pp. 853–870, 2018, doi:10.1016/j.compeleceng.2018.01.002.

[31] P. Porambage, J. Okwuibe, M. Liyanage, M. Ylianttila, and T. Taleb, "Survey on multi-access edge computing for Internet of Things realization," *IEEE Commun. Surv. Tutorials*, vol. 20, no. 4, pp. 2961–2991, 2018. doi:10.1109/COMST.2018.2849509.

[32] M. W. Akhtar, S. A. Hassan, R. Ghaffar, H. Jung, S. Garg, and M. S. Hossain, "The shift to 6G communications: Vision and requirements," *Human-centric Comput. Inf. Sci.*, vol. 10, no. 1, pp. 1–27, 2020, doi:10.1186/s13673-020-00258-2.

[33] M. Mukherjee, L. Shu, and D. Wang, "Survey of fog computing: Fundamental, network applications, and research challenges," *IEEE Commun. Surv. Tutorials*, vol. 20, no. 3, pp. 1826–1857, 2018, doi:10.1109/COMST.2018.2814571.

[34] A. Mughees, M. Tahir, M. A. Sheikh, and A. Ahad, "Towards energy efficient 5G networks using machine learning: Taxonomy, research challenges, and future research directions," *IEEE Access*, vol. 8, pp. 187498–187522, 2020, doi:10.1109/ACCESS.2020.3029903.

[35] T. Taleb, K. Samdanis, B. Mada, H. Flinck, S. Dutta, and D. Sabella, "On multi-access edge computing: A survey of the emerging 5G network edge cloud architecture and orchestration," *IEEE Commun. Surv. Tutorials*, vol. 19, no. 3, pp. 1657–1681, 2017, doi:10.1109/COMST.2017.2705720.

[36] S. R. Bader, M. Maleshkova, and S. Lohmann, "Structuring reference architectures for the industrial Internet of Things," *Futur. Internet*, vol. 11, no. 7, pp. 1–23, 2019, doi:10.3390/fi11070151.

[37] Z. Rasheed *et al.*, "Brain tumor classification from MRI using image enhancement and convolutional neural network techniques," *Brain Sci.*, vol. 13, no. 1320, p. 22, 2023, doi:10.3390/brainsci13091320.

[38] Q. V. Pham *et al.*, "A survey of multi-access edge computing in 5G and beyond: Fundamentals, technology integration, and state-of-the-art," *IEEE Access*, vol. 8, pp. 116974–117017, 2020. doi:10.1109/ACCESS.2020.3001277.

[39] S. Waleed, I. Ullah, W. U. Khan, A. U. Rehman, T. Rahman, and S. Li, "Resource allocation of 5G network by exploiting particle swarm optimization," *Iran J. Comput. Sci.*, vol. 4, no. 3, pp. 211–219, 2021, doi:10.1007/s42044-021-00091-5.

[40] I. Khan, Q. Wu, I. Ullah, S. U. Rahman, H. Ullah, and K. Zhang, "Designed circularly polarized two-port microstrip MIMO antenna for WLAN applications," *Appl. Sci.*, vol. 12, no. 3, p. 1068, 2022, doi:10.3390/app12031068.

[41] R. Khan et al., "3D convolutional neural networks based automatic modulation classification in the presence of channel noise," *IET Commun.*, vol. 16, no. 5, pp. 497–509, 2022, doi:10.1049/cmu2.12269.

[42] H. A. Belghol and A. Idrissi, "MEC towards 5G : A survey of concepts, use cases, location trade-offs MEC towards 5G : A survey of concepts, use cases, location trade-offs," August, vol. 5, no. 4, pp. 1–10, 2017, doi:10.14738/tmlai.54.3215.

[43] T. X. Tran, A. Hajisami, P. Pandey, and D. Pompili, "Collaborative mobile edge computing in 5G networks: New paradigms, scenarios, and challenges," *IEEE Commun. Mag.*, vol. 55, no. 4, pp. 54–61, 2017, doi:10.1109/MCOM.2017.1600863.

[44] T. Mazhar et al., "The role of ML, AI and 5G technology in smart energy and smart building management," *Electronics*, vol. 11, no. 23, pp. 1–21, 2022, doi:10.3390/electronics11233960.

[45] A. Gohar and G. Nencioni, "The role of 5g technologies in a smart city: The case for intelligent transportation system," *Sustainability (Switzerland)*, vol. 13, no. 9. pp. 1–24, 2021, doi:10.3390/su13095188.

[46] M. Asif et al., "Reduced-complexity LDPC decoding for next-generation IoT networks," *Wirel. Commun. Mob. Comput.*, vol. 2021, pp. 1–10, 2021, doi:10.1155/2021/2029560.

[47] W. U. Khan, N. Imtiaz, and I. Ullah, "Joint optimization of NOMA -enabled backscatter communications for beyond 5G IoT networks," *Internet Technol. Lett.*, vol. 4, no. 2, p. e265, 2021, doi:10.1002/itl2.265.

[48] M. H. Ur Rehman, P. P. Jayaraman, S. Ur Rehman Malik, A. Ur Rehman Khan, and M. M. Gaber, "RedEdge: A novel architecture for big data processing in mobile edge computing environments," *J. Sens. Actuator Netw.*, vol. 6, no. 3, pp. 1–22, 2017, doi:10.3390/jsan6030017.

[49] M. Chiang, S. Ha, I. Chih-Lin, F. Risso, and T. Zhang, "Clarifying fog computing and networking: 10 Questions and answers," *IEEE Commun. Mag.*, vol. 55, no. 4. pp. 18–20, 2017. doi:10.1109/MCOM.2017.7901470.

[50] Sumit et al., "Energy saving implementation in hydraulic press using industrial Internet of Things (IIoT)," *Electronics*, vol. 11, no. 23, p. 4061, 2022, doi:10.3390/electronics11234061.

[51] F. Wahab, I. Ullah, A. Shah, R. A. Khan, A. Choi, and M. S. Anwar, "Design and implementation of real-time object detection system based on single-shoot detector and OpenCV," *Front. Psychol.*, vol. 13, no. November, pp. 1–17, 2022, doi:10.3389/fpsyg.2022.1039645.

[52] A. Iqbal et al., "Bidirectional CPW fed quad-band DRA for WLAN/WiMAX applications," *Wirel. Commun. Mob. Comput.*, vol. 2022, 9 pages, 2022, doi:10.1155/2022/3763555.

[53] TSGS, "System architecture for the 5G system (5GS) (3GPP TS 23.501 version 16.6.0 Release 16)," vol. 0. 2020.

[54] National Academies of Sciences, Engineering, and Medicine. "The Transformational Impact of 5G: Proceedings of a Workshop–in Brief (2019)," 2019.

[55] I. Ullah, J. Chen, X. Su, C. Esposito, and C. Choi, "Localization and detection of targets in underwater wireless sensor using distance and angle based algorithms," *IEEE Access*, vol. 7, pp. 45693–45704, 2019, doi:10.1109/ACCESS.2019.2909133.

[56] H. U. Khan, M. Sohail, F. Ali, S. Nazir, Y. Y. Ghadi, and I. Ullah, "Prioritizing the multi-criterial features based on comparative approaches for enhancing security of IoT devices," *Phys. Commun.*, vol. 59, p. 102084, 2023, doi:10.1016/j.phycom.2023.102084.

[57] R. Pal, D. Adhikari, M. B. Bin Heyat, I. Ullah, and Z. You, "Yoga meets intelligent Internet of Things: Recent challenges and future directions," *Bioengineering*, vol. 10, no. 4, pp. 1–25, 2023, doi:10.3390/bioengineering10040459.

[58] A. Shah, B. Ali, M. Habib, J. Frnda, I. Ullah, and M. Shahid Anwar, "An ensemble face recognition mechanism based on three-way decisions," *J. King Saud Univ. - Comput. Inf. Sci.*, vol. 35, no. 4, pp. 196–208, 2023, doi:10.1016/j.jksuci.2023.03.016.

[59] S. M. Shah et al., "Advancements in neighboring-based energy-efficient routing protocol (NBEER) for underwater wireless sensor networks," *Sensors*, vol. 23, no. 13, p. 6025, 2023, doi:10.3390/s23136025.

[60] H. U. Khan, A. Hussain, S. Nazir, F. Ali, M. Z. Khan, and I. Ullah, "A service-efficient proxy mobile IPv6 extension for IoT domain," *Information*, vol. 14, no. 8, p. 459, 2023, doi:10.3390/info14080459.

[61] J. Rao and S. Vrzic, "Packet duplication for URLLC in 5G: Architectural enhancements and performance analysis," *IEEE Netw.*, vol. 32, no. 2, pp. 32–40, 2018, doi:10.1109/MNET.2018.1700227.

[62] J. Gubbi, R. Buyya, S. Marusic, and M. Palaniswami, "Internet of Things (IoT): A vision, architectural elements, and future directions," *Futur. Gener. Comput. Syst.*, vol. 29, no. 7, pp. 1645–1660, 2013, doi:10.1016/j.future.2013.01.010.

Index

5G 4, 14, 24, 35, 40, 81–82, 86, 138, 214, 216, 221, 226, 228
6G 14, 214

AI 3–5, 20, 23–25, 31–32, 37, 41–43, 57–60, 110, 202
AIoT 4, 13–14, 41–45, 50–51

big data 25, 41, 49, 80–82, 92–101, 103–107, 168, 219

cloud computing 37, 81, 83–84, 174, 178, 180–181, 193–200, 215–219, 221–225
communication systems 31, 37, 62, 64, 66, 74, 215
cyber ethics 110, 112, 116–118

data science 6, 9, 42, 73–76, 84–88, 92, 107
deep learning 4, 19, 21, 40, 47, 74, 76, 93, 101–103, 113, 134, 161
digital twins 137–138, 141, 146–149

edge computing 14, 25, 75, 80–81, 83, 181, 214–220

fog computing 83, 216, 218–219, 222
future world 110, 112, 117, 127

IoT 3–4, 6–14, 31–46, 133–134, 136–138, 144, 160, 202–203, 224, 226
IoV 168, 182, 205–211, 227

M2M 35, 39, 64, 226, 229
machine learning 3, 36, 62, 64–65, 74–76, 78–81, 84, 102, 152, 161, 166, 202
Metaverse 133–134, 136–150

NLP 21, 60, 62, 93–94, 100–101, 103–104, 106, 115

reinforcement learning 8, 19–20, 23, 65
routing 111, 115, 194–195, 197, 206, 208–211
RPL 206, 210–211

SDN 193–199, 201–204, 223–224
sensor 10, 20, 31–32, 39–40, 74, 78, 112, 116, 136, 181, 184, 206, 227
SIoT 160–169, 171–172, 174–186
smart cities 4, 13, 18, 20–21, 31, 77–78, 87, 136, 139, 141, 153, 225–227, 229
social networks 98, 141, 160, 166, 168, 170, 182, 184

V2V 211
virtual reality 66, 134, 151, 215, 221, 228

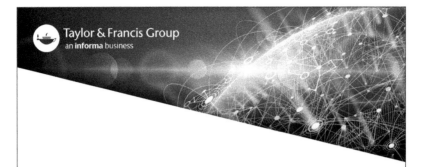